THE CHEMISTRY OF FOODS.

WITH MICROSCOPIC ILLUSTRATIONS.

BY

JAMES BELL, Ph.D. &c.,

Principal of the Somerset House Laboratory, and Vice-President of the Institute of Chemistry, etc.

Part I.

TEA, COFFEE, COCOA, SUGAR, ETC.

Published for the Committee of Council on Education
BY
CHAPMAN AND HALL, Limited,
11, HENRIETTA STREET, COVENT GARDEN.
1881.

PREFACE.

This Work is intended partly as an aid to those who may desire to carefully examine the illustrations of Food Adulteration in the Bethnal Green Branch of the South Kensington Museum, and partly as a contribution to the already published knowledge of the composition and analysis of articles of Food.

The analyses, with the exception of those under the head of Coffee, which were made for the Inland Revenue Department in 1852, are almost entirely new, and have been made with great care for departmental purposes.

It is intended that this work shall be followed by Parts II. and III., which will treat of the analysis and composition of other foods.

It can scarcely be expected that a work like this, prepared amidst the labours of an important Government department, will be found entirely free from errors, but it is hoped that these are not numerous.

JAMES BELL.

Laboratory Department,
  Somerset House,
    August, 1881.

PART I.

CONTENTS.

	PAGE
TEA	1
COFFEE	40
COCOA	72
SUGAR	97
HONEY	115
APPENDIX—Table for converting Degrees of Centigrade Thermometer into Degrees of Fahrenheit's Scale	117
Comparison of French and English Measures of Weight	118
INDEX	119

TEA.

Botanical Origin.—The tea of commerce is derived from a plant named *Thea sinensis*, which is an evergreen, and closely allied to the well-known genus *Camellia*. In cultivation, the tea plant is generally found from 3 to 6 feet high, but if allowed to attain to its full growth it reaches to a height of between 30 and 40 feet, and possesses a stem more than 1 foot in diameter. It is a native of China, Japan, and the northern parts of Eastern India; but its cultivation has been successfully introduced into some parts of British India, and it has been grown with a fair amount of success in Ceylon, Brazil, and Carolina.

At one time it was thought that black and green teas were prepared from the leaves of two different plants, named respectively *Thea bohea* and *Thea viridis*; but it is now known that one plant, *Thea sinensis*, is the source of both kinds, and that black or green tea can be prepared at pleasure from the same leaves, the difference depending entirely upon the process followed in the manufacture.

Description.—The tea of commerce consists of the prepared leaves of the tea plant, but in most samples there are present portions of the young branches and flower buds. The leaves are sometimes 2 inches long and 1 inch wide; usually, however, they are much smaller, though the full-grown leaf measures from 5 to 6 inches in length.

The leaves are gathered four times during the year, the first gathering being made early in the spring, and the following three at intervals of about six weeks between each. The tea prepared from the first gathering is most delicate in colour and flavour, contains the smallest proportion of woody fibre, and is reckoned the best in quality. The quality of the tea depends also upon the age of the tree as well as upon the age of the leaf, the finest teas being produced from the young leaves of young plants; whilst old leaves, and the leaves of old wood, are deficient both in flavour and extract.

In the first stage of the preparation of green tea, the leaves are gently heated for a few minutes to render them soft and flaccid, and after being removed from the pans they are rolled by the hand on a wooden table. They are again placed in the drying-pans, and the leaves are carefully kept in motion until sufficiently dry, the operation being proceeded with as quickly as possible to conserve the green colour and prevent fermentation.

The leaves intended for black tea are thrown into heaps to undergo fermentation, and at the expiration of a few hours are tossed about until they become quite soft, when they are rolled into balls by a peculiar movement and strong pressure of the hands. They are next exposed to the air for some hours, and then alternately dried and rolled two or three times, and finally dried over a charcoal fire.

It would appear that this process is not uniformly followed, as in some instances the leaves are partly withered by exposure to the sun for two hours, and then tossed and beaten by the hands for some time until they become flaccid. This is repeated two or three times at intervals of about half an hour. The fermentative action appears to proceed during this part of the process. The leaves are next heated in an iron pan for a short time, and then rolled into balls, by which operation some of the juice is expressed. Having been half dried over a charcoal fire, they are removed to an open basket and allowed to remain until next day, when the drying is again proceeded with, and continued with various precautions, until the tea has attained sufficient crispness to be considered ready for the market. The dried tea is then carefully sorted, by sifting it through sieves of different-sized meshes, and by hand-picking. In this way it is divided into parcels of uniformly-sized leaves, and at the same time every unsightly and imperfectly-dried leaf is removed.

Some teas are scented in order to impart to them an agreeable flavour. The flavour is communicated by placing the leaves in contact with the flowers of plants possessing an aromatic odour, such, for example, as the flowers of *Olea fragrans*, which are used in the preparation of scented Pekoes.

The two great classes of tea, green and black, are each subdivided into a variety of kinds which are known in commerce by particular names. In green teas we have Gunpowder, Hyson, Young Hyson, Imperial, and Twankay; and in black teas, Congou, Kaisow, Moning, Souchong, Oolong, and Assam.

A tea is pronounced of good quality when it possesses delicacy and fulness of flavour with a certain amount of body, and its value in the market is determined by the extent to which it possesses these characters.

Tea does not belong to the class of nutritive substances, but is chiefly prized on account of its refreshing and stimulating properties, the most obvious of which are those of engendering activity of thought, driving away sleep, and stimulating to greater muscular exertion. These effects, however, are not produced in the same degree in different persons; what in one produces simply a soothing sensation, gives rise, perhaps, in another, to a high state of excitability. Taken in excess, it is said to produce giddiness and nervous trembling.

History.—Little is known of the early history of the use of the

in countries other than China and India. The comparative failure of tea-growing in some countries has been attributed not so much to climate and soil as to the want of skilled labour.

According to the "Encyclopædia Britannica," the annual consumption per head of the population in the United Kingdom is 36 ounces. In England the consumption is 40 ounces, in Scotland 35 ounces, and in Ireland 23 ounces per head.

CHEMICAL COMPOSITION.

Tea has been the subject of numerous investigations as to the nature and quantity of its several constituents. There has been a certain measure of agreement in the statements which treat of the kind of substances which form the bulk of the tea leaf, but a very great diversity exists in the results stated to have been obtained by different chemists with regard to the quantities in which these are present. This is especially true if we include the earlier investigations on the subject, the results of which, with regard to some of both the organic and mineral constituents, have not been confirmed by the more recent researches. The differences may partly arise from the various meanings which it is possible to attach to some of the terms in which the analyses are stated; such as, for example, "Extractive," "Gum," "Sugar," "Tannin," and "Albumin." It may be said that the only organic constituent of tea which has been completely isolated and identified is the alkaloid theine, and this is no doubt due to the facility with which it crystallizes.

The organic substances found to exist in tea are a volatile oil, to which much of the characteristic odour of tea is due, theine, tannin, an albuminous body, gum or dextrin, pectin, cellulose, chlorophyll, and resin.

The following results were obtained from the analysis of a Congou Tea at 2s. 10d. and Young Hyson at 3s. per lb. They were selected as being fair representatives of black and green teas.

	Congou.	Young Hyson.
Moisture	8·20	5·96
Theine	3·24	2·33
Albumin, insoluble	17·20	16·83
,, soluble	·70	·80
Extractive by alcohol, containing nitrogenous matter	6·79	7·05
Dextrin or Gum	—	·50
Pectin and Pectic Acid	2·60	3·22
Tannin	16·40	27·14
Chlorophyll and Resin	4·60	4·20
Cellulose	34·00	25·90
Ash	6·27	6·07
	100·00	100·00

Oil of Tea.—The essential oil of tea is present in very small quantity. It has a specific gravity less than water, is of a yellowish colour, and readily passes into the form of a resin by exposure to the air. It possesses the peculiar taste and smell of tea, and has very potent stimulating properties. Taken in rather large quantities, the oil is said to produce headache and giddiness.

The peculiar odour of tea is mostly developed during the process of manufacture. It is more than doubtful whether it arises solely from a definite body pre-existing in the tea, as we have found that the flavour of black tea was produced, by heating for some time to a temperature of 212° F. (100° C.), a portion of an extract of green tea from which the oil or resinous matter had been removed.

Theine, $C_8 H_{10} N_4 O_2$.—This is the alkaloid of tea. The proportion in which it is present was for some time greatly underestimated by chemists. More recent analyses, however, show a greater quantity. Stenhouse has found from 1·05 to 4·1 per cent.; Peligot, from 2·3 to 4·1 per cent.

In some recent analyses made by ourselves we obtained the following amounts of theine from 100 grains of the tea dried at 212° F. (100° C.):

Congou, low	2·78 grains.
Do. fine	3·12 ,,
Hyson	2·24 ,,
Souchong	2·97 ,,
Moning	2·93 ,,
Assam	3·42 ,,
Gunpowder	2·72 ,,

Theine is very rich in nitrogen, of which it contains nearly 29 per cent. Albumin and similar substances contain only from 15 to 16 per cent.

It is to theine, chiefly, that the beneficial and stimulating properties of tea are ascribed, aided, no doubt, by the peculiar volatile principle present in the prepared leaf. Theine exists in combination with tannin in tea, and it is an impure compound of these substances which precipitates on allowing a rather concentrated hot-water solution of tea to cool. Theine crystallizes from water in the form of long needles of a white and silky lustre, containing one atom of water of crystallization. It sublimes at 365° F. (185° C.), and an attempt has been made to take advantage of this property to estimate the amount of theine in tea, but without any marked success. It dissolves rather freely in hot water, less so in cold water and alcohol, and with still greater difficulty in ether. It is altered by boiling with nitric acid, the product forming, with vapour of ammonia, a coloured substance which resembles murexide, produced in a similar way from uric acid.

Albumin or Vegetable Casein.—This substance exists almost entirely in the insoluble form in tea. A small quantity is dissolved out with water, but the amount is less than 1 per cent. Like ordinary casein and coagulated albumin, it is dissolved by alkalis; but its separation by this means from the cellulose of the leaf is unsatisfactory.

The cellulose of tea is readily acted on by the fixed alkalis, so that the albumin can be only partially recovered in an impure state. The amount of this substance may be more accurately determined by thoroughly exhausting the leaf, first with alcohol and then with water, and estimating the nitrogen in the portion of the leaf remaining insoluble, reckoning the quantity so obtained as being all derived from albumin.

When the nitrogen, associated with the cellulose in the form of vegetable albumin, is deducted from the total amount of nitrogen found in the leaf, a quantity remains which cannot be accounted for by any proportion of theine which has as yet been fairly obtained from tea. The alcoholic extract, therefore, either contains a larger amount of theine than has been recovered from it, or there is present a quantity of another and undetermined nitrogenous substance.

Gum or Dextrin.—Substances under the indefinite term of "gum" are stated by chemists to be present to the extent of from 5 to 9 per cent. We have found, however, in samples of black and green teas, the analyses of which are given above, that dextrin, arabin, or similar gum, convertible into sugar by sulphuric acid, was practically absent. It is true that about ½ per cent. of a gum corresponding to dextrin was found in the green tea; but unless the Chinese are exonerated from the suspicion of using such a gum in making up green teas, it is open to question whether even this small proportion is natural to the leaf.

Pectin, etc.—The characteristic gummy matter of tea appears to be pectin and pectic acid. It is obtained in considerable purity from the water extract after the tea has been well exhausted by alcohol. It is precipitated by alcohol in presence of hydrochloric acid as a transparent jelly, the reactions of which, on subsequent treatment with acids and alkalis, are those of pectin and pectic acid.

Sugar.—Neither of the two descriptions of tea gave any indication of sugar. The tannin of the green tea gave, after boiling with a little dilute mineral acid, 1·33 per cent. of glucose, indicating that a portion of it existed as a glucoside. Under similar conditions the tannin of the black tea gave no sugar.

Tannin.—This is the most abundant substance found in the soluble part of the tea-leaf. Although in some degree it answers to ordinary gallo-tannic acid in its reactions, yet, from its instability and the modifications it undergoes under chemical treatment, we are inclined to the opinion that it differs from that acid in some important respects.

Chlorophyll and Resin.—Tea contains a small quantity of certain substances soluble in ether and benzol, and insoluble in water. These chiefly consist of chlorophyll and resinous bodies. It is probable that the amount obtained from tea is greater than what was originally present in the leaf, as some of the tannin and other constituents are liable to be changed by oxidation into a resinous-like substance.

Cellulose.—The cellulose or woody fibre, which is insoluble in water, forms a considerable proportion of the tea-leaf. After extracting all the soluble constituents of the tea with water, there are left associated with the cellulose nearly all the albumin, part of the ash, and a little of the colouring matter. These cannot be well separated without loss of cellulose, the estimation of which has consequently to be determined by difference.

Ash.—The following table exhibits the composition of the ash of seven descriptions of tea, including two qualities of Congou.

CONSTITUENTS OF THE ASH OF TEA.

Percentage of	Congou (low).	Congou (fine).	Hyson.	Souchong.	Moning.	Assam.	Gunpowder (fine).
Total Ash on Dry Tea	6·10	6·94	6·46	5·99	8·29	6·49	6·67
Sand	3·08	8·51	2·17	1·51	13·37	3·72	5·66
Silica	6·35	9·27	5·93	3·77	9·47	2·51	6·52
Chlorine	1·06	1·07	1·12	1·01	·99	·97	1·11
Potassium, to satisfy Chlorine	1·16	1·17	1·23	1·11	1·09	1·07	1·22
Potash ... estimated as K_2O	34·38	28·87	35·66	34·29	26·83	37·71	30·69
Soda ,, Na_2O	·62	1·07	·80	·34	·50	·97	1·27
Oxide of Iron ,, FeO	2·82	·84	1·12	1·68	2·23	·57	1·43
Alumina ,, Al_2O_3	5·55	3·42	2·73	4·19	4·52	1·54	2·70
Oxide of Manganese ,, Mn_3O_4	1·68	1·37	1·93	1·59	1·49	2·11	1·92
Lime ,, CaO	8·82	8·74	9·54	8·98	9·04	8·58	8·19
Magnesia ,, MgO	2·12	4·87	4·65	3·19	2·42	6·48	6·52
Phosphoric Anhydride ,, P_2O_5	14·11	14·68	14·11	18·54	12·69	14·74	16·15
Sulphuric ,, SO_3	6·52	6·54	6·34	6·38	5·39	5·83	6·68
Carbonic ,, CO_2	11·73	9·58	12·67	13·42	9·97	13·20	9·94
	100·00	100·00	100·00	100·00	100·00	100·00	100·00

In these analyses of tea-ash it will be observed that soda is present in uniformly low quantities. The percentage of iron is lower than that given in some published analyses of tea-ash, but no account is given therein of the alumina which appears to be a constant constituent of the ash, and which was probably included with the iron. The presence of sulphuric acid appears to have been disregarded by most chemists. We find that it is present in remarkably constant percentages.

The sample of "Moning" gives a high proportion of total ash (8·29), arising from sand and silica. It will be seen on page 28 that a second sample shows only 6·88 per cent.

MICROSCOPIC STRUCTURE.

The leaves of the tea plant vary in size, but seldom exceed

FIG. I.—TEA.

2 inches in length and 1 inch in breadth (Fig. 1, A and B).

TEA. 13

The form of the leaf is that of an ellipse, terminating in a slightly emarginate apex. The margin is dentate, each tooth supporting a

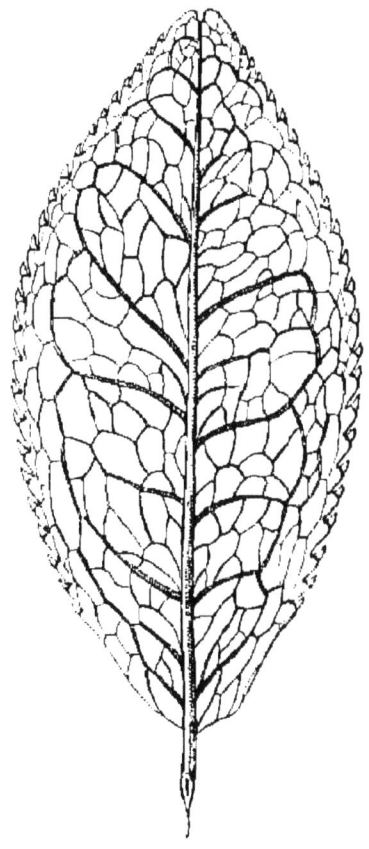

FIG. 2.—TEA.

short caducous spine. The spines are best seen on the fresh green leaf, of which an enlarged representation is shown in Fig. 2.

The principal veins proceed from different points of the midrib, ending in curvatures within the margin, and forming on each side of the midrib a row of oblong meshes or loops.

The epidermis of the under surface of the leaf, seen under the

FIG. 3.—SKIN OF TEA-LEAF.

microscope, consists of well-marked sinuous cells with numerous oval stomates, and a few simple unicellular tortuous hairs (Fig. 3). The skin of the upper surface is similar in structure, but the cells are smaller, and there are no stomates.

The interior of the leaf is made up of fibro-vascular tissue, like that composing the midrib and veins, surrounded on all sides by a network of round or oval cells filled with chlorophyll grains (Fig. 4, A). Scattered amongst these cells are peculiar branched bodies which have been sometimes called "branched and spinous hairs" (Fig. 4, c), but which are, in reality, branched, thick-walled cells. There are also

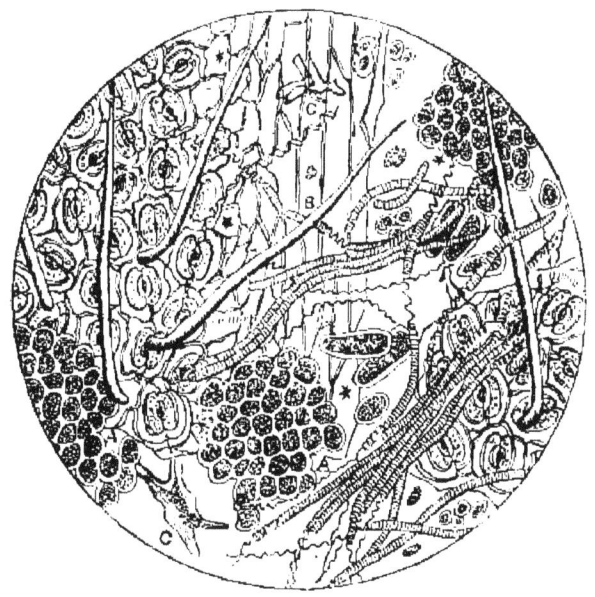

FIG. 4.—GROUND TEA.

present in the leaf numerous sphœraphides and other plant crystals.

Chief Characteristics.—These are the looped or meshed venation of the leaf, the dentate margin with the short spines, the slightly emarginate apex, the numerous stomates, the unicellular hairs, and the peculiar branched cells referred to.

ANALYSIS.

Although, in the analysis of tea, the more characteristic principles have chiefly engaged the attention of chemists, a more or less complete analysis of the leaf has been attempted. Various methods appear to have been adopted to arrive at an accurate estimation of the several constituents; but, from the very different results obtained, some of them must obviously be incorrect.

Instead of making a water extract only, we have first exhausted the tea as completely as possible with alcohol of 70 per cent., and then with water, keeping the extracts apart and examining them separately. The result of following this process is that a larger total extractive is obtained than by water only, nearly all the tannin and soluble nitrogenous matter being found in the alcoholic extract, while at the same time the constituents soluble in water can be more readily dealt with.

Oil.—This is determined by distilling about 100 grains of the tea in presence of water. The distillate possesses a strong odour of tea, and contains a little chlorophyll and resinous matter mechanically carried over, which should be removed by filtration. The filtrate is saturated with chloride of calcium and shaken with ether in a separator. The upper layer of liquid is then drawn off into a tared beaker, and the ether evaporated at a temperature of about 80° F. (26·6° C.). The weight of the beaker and oil is then ascertained.

Theine.—One of the two following methods is usually adopted for the estimation of theine in tea:

First Method.— One hundred grains of dry and powdered tea are boiled for several minutes with an equal weight of calcined magnesia and 8 ounces of strong alcohol, and filtered. The boiling is repeated with a like amount of alcohol, and subsequently three times with distilled water, filtering after each operation. The alcohol is recovered by distillation from the united alcoholic filtrates, and the residue, after addition of water, is filtered to remove colouring matter thrown out of solution. A further amount of colouring matter is got rid of by evaporating this filtrate to dryness, and exhausting the residue with boiling water. The resulting solution is added to the aqueous extracts obtained directly from the tea, and the whole evaporated to dryness with the addition of a further small quantity of magnesia, and completely exhausted with pure hot benzol. On distillation, or evaporation of the benzol, the theine is left in a crystalline and almost colourless state.

Second Method.—One hundred grains of dried tea are reduced to a powder, and thoroughly exhausted with boiling water. To the extract, subacetate of lead is added as long as a precipitate appears. The precipitate is then removed by filtration, and the lead in the filtrate got rid of by means of sulphuretted hydrogen. The filtrate from the sulphide of lead is gently evaporated to a small bulk, and left for twenty-four hours, when the theine will be found to crystallize out in an impure state. The crystals are removed from the mother liquor, pressed between bibulous paper, dissolved in a little water, and left to re-crystallize. The mother liquor still contains a little theine, which some chemists have endeavoured to estimate by a standard solution of tannic acid. In the course of our experiments we have found the first of these methods to yield the best results.

Tannin.—The amount of tannin in tea is sometimes roughly estimated by weighing the precipitate occasioned by a solution of gelatine and alum in a given quantity of tea solution and reckoning 40 per cent. of the dried product as tannin. By a second method 4 grams of gelatine and 2 grams of alum are dissolved in warm water and made up to 1 litre; then 1 grain of the finest tannic acid is dissolved in 40 cubic centimetres of water and 10 cubic centimetres of a saturated solution of common salt, added along with a little ground glass. The gelatine solution is introduced into a burette, and run into the solution of tannic acid until no further precipitate forms. The number of cubic centimetres is read off and noted.

Twenty-five grains of the powdered tea are next exhausted with boiling water, the solution filtered and made up to 200 cubic centimetres. 40 cubic centimetres of the solution (= 5 grains of tea) are transferred to a beaker, and solution of salt and some ground glass added as above. The process is now conducted as in the test solution. There appears to be an advantage in making the liquid in the beaker specifically heavier than the gelatine solution, as the latter, when carefully dropped into the beaker, remains for a short time on the surface, and thus allows any formation of precipitate to be more easily discerned.

Suppose that the one grain of tannic acid requires 14 cubic centimetres of gelatine solution, and 40 cubic centimetres of tea solution require 9·8 cubic centimetres of the gelatine, then 14 : 1 : 9·8 × 20 = 14 per cent. of tannic acid in the sample of tea.

The following method has been proposed by Mr. Allen. A solution of acetate of lead in distilled water is prepared of a strength equal to 5 grams in the litre. 1-10th gram of pure tannic acid is dissolved in 100 cubic centimetres of water, and introduced into a burette; 10 cubic centimetres of the lead solution are measured into a beaker, and diluted with 90 cubic centimetres of water. This solution is made hot, and the tannic acid solution carefully dropped into it until a little of the filtered liquid gives a pink colouration with a drop of an ammoniacal solution of ferricyanide of potassium.

The tea solution is prepared by exhausting 2 grams of the powdered tea with boiling water, and making up to 250 cubic centimetres. This solution is tested in the same way, and the calculation of the percentage of tannin made from the two results. It is evident that some precautions are necessary in applying the test, as there is no provision made here for the precipitation of the lead by the alkaline phosphates and other substances present in the solution which are thrown down by a salt of lead.

Another method by Löwenthal, with modifications by Estcourt, has been suggested. For this test various solutions are required:

1. Solution of permanganate of potassium made by dissolving 1·66 grams of the salt in a litre of water.

2. Solution of pure sulphate of indigo: 30 grams dissolved in water and made up to a litre.

3. Solution of pure tannic acid: 2 grams in 1 litre of distilled water.

4. Dilute sulphuric acid: 200 cubic centimetres of concentrated sulphuric acid diluted to 1 litre, or 1 in 5.

5. Solution of pure gelatine: 25 grams of best glue dissolved in warm water, and made up to 1 litre with a saturated solution of common salt.

In the application of the test, the tannic acid, indigo, and other matters in the tea solution are oxidised by the permanganate, and the end of the process is shown by the change of the blue colour to green and then to a pale yellow.

The first point to be ascertained is the number of cubic centimetres of permanganate solution required to decolourise 20 cubic centimetres of the indigo solution. For this purpose 20 cubic centimetres of the solution are diluted with 800 cubic centimetres of distilled water and 10 cubic centimetres of the sulphuric acid solution added. The permanganate solution is run in from a burette until the blue colour completely disappears— the end of the process being carefully observed. The same quantity of indigo solution is again taken, with the addition of 10 cubic centimetres of the tannin solution, and the permanganate solution added as before. The difference in the two results is due to the tannic acid. It is desirable that the experiments should occupy the same time—about 6 minutes—and the solution should be kept vigorously stirred during the addition of the permanganate.

Two grams of tea are now exhausted with boiling water and the extract made up to 250 cubic centimetres. 10 or 15 cubic centimetres of the tea solution are tested in the manner just described; the result represents the total oxidisable matter in the tea. To ascertain how much is due to the tannin, it has been proposed to precipitate the tannic acid in a given quantity of the tea extract with the gelatine solution, filter off a proportionate quantity, and test its oxidisable value by the standard permanganate. This latter indication is deducted from the amount obtained in the first experiment with the solution of tea, and the difference gives the number of cubic centimetres required to oxidise the tannic acid in the tea.

Albumin or Vegetable Casein.—Six grams of finely powdered tea are digested with 3 ounces of alcohol of 70 per cent. at a temperature of 140° F. (60° C.) for fifteen minutes and filtered, the process being repeated three or four times with successive quantities of alcohol. The residue is next treated with water, and, after having been digested in a similar way, as in the first operation, is thrown on a filter and well washed with hot water.

The insoluble portion which remains on the filter is dried and weighed; and, in order to determine the amount of nitrogen present, about ·7 gram is submitted to combustion with copper oxide, and the proportion of albumin calculated by multiplying the amount of nitrogen obtained by the factor 6·3—the relation by weight existing between albuminous substances and the nitrogen they contain being as 6.3 : 1.

As already pointed out, a small quantity of albumin is precipitated by boiling the water extract. This precipitate is weighed and the nitrogen ascertained in the same way as above.

Pectin and Pectic Acid.—The filtrate from the precipitated albumin is evaporated to a small bulk, and after the addition of a few drops of hydrochloric acid, the pectin and pectic acid are precipitated by the addition of 200 cubic centimetres of alcohol of 90 per cent. The precipitate, which appears as a transparent gelatinous mass, is separated by filtration, re-dissolved in a little water, and again precipitated as before. The precipitate when well washed is dried and weighed on a tared filter. A portion is then ignited to ascertain the amount of ash present, and the remainder is boiled for four hours with water acidulated with 10 drops of sulphuric acid to convert any dextrin or similar substance into glucose. The weight of ash and dextrin, if any, is deducted from the weight of the alcoholic precipitate previously ascertained, the difference being regarded as the percentage of pectin and pectic acid present.

Dextrin.—The proportion of dextrin or gum is ascertained from the amount of sugar produced by boiling the alcoholic precipitate with dilute sulphuric acid, as in the above process for the estimation of pectin. The quantity of sugar formed is determined by an alkaline copper solution, as described on page 107, from which result the dextrin can be readily calculated: 100 parts of glucose or 95 parts of cane-sugar being equal to 90 parts of dextrin.

Cellulose.—The percentage of cellulose cannot be determined by direct analysis, and the process followed is usually an indirect one. The amount of cellulose is represented by the portion insoluble in alcohol and water, less the mineral matter, and insoluble albumin. The ash is obtained by igniting a known weight of the dry residue, and the albumin, as before stated, by combustion with copper oxide. The difference is the percentage of cellulose and insoluble colouring matter present in the tea.

Chlorophyll and Resin.—Fifty grains of the dry and finely-powdered tea are left in contact with ether for twenty-four hours. The ether is passed through a filter, and the tea-powder washed with warm ether. The filtrate is evaporated to dryness, and treated with hot water to dissolve out any substances soluble therein. The portion insoluble in water is then dried and weighed.

The powder left after treatment with ether is shaken up with alcohol for some time, the alcoholic extract is filtered and gently evaporated to dryness. The residue is exhausted with benzol, and, after filtering and evaporating the benzol, the extract is treated with boiling water, as in the former part of the process. After removing the water, the resinous mass is dried and weighed. To this result is added the quantity of extract obtained by ether; the sum multiplied by 2 gives the percentage shown as "chlorophyll and resin."

ADULTERATION.

At one time there were probably few articles so generally adulterated as tea, especially when its price was such as to hold out strong inducement for the manufacture and sale of a spurious commodity. The sophistication was carried on both in China and in this country; but for many years past the adulteration of tea has been effected almost entirely before its importation.

The spurious teas manufactured in this country were composed sometimes of exhausted tea-leaves, and sometimes of the leaves of other plants, such as the elder, the sloe, and the willow. The leaves, when made up to represent green tea, were generally prepared with some of the following substances: viz. gum, Dutch pink, Prussian blue, indigo, carbonate of magnesia, French chalk, and sulphate of lime; and, when made to represent black tea, the leaves were slightly coloured with Dutch pink to impart a bloom.

The sophistication of teas imported from China has most frequently consisted of partially exhausted leaves, and of leaves made up with a large proportion of sand or broken quartz, by the aid of a little gum, and then skilfully rolled and re-dried so as to resemble ordinary commercial teas. Plumbago was generally employed to give uniformity of colour to black, and a mixture of Prussian blue and soapstone to green teas.

Happily, at the present time, a great change has taken place for the better in the quality of the teas found in commerce and supplied to the public. It is now of very seldom occurrence that quartz or foreign leaves are found mixed with tea, or that colouring matters to the same extent, or in the same variety as formerly, are discovered to have been used in facing tea.

The adulterants of tea may be classed under three heads:

1st. Those substances which can be detected by their physical properties—such as foreign leaves, quartz, excess of sand, and certain colouring matters.

2nd. Those which can be distinguished by their chemical properties—for example, Prussian blue, clay, soapstone, gum, rice-water, etc.

3rd. Partially exhausted leaves.

The quartz and sand are usually made up with the tea when in a soft and flaccid condition, and are ingeniously concealed in the nodules formed of the leaves. Sometimes irregular grains of quartz have been simply coloured with plumbago or other substance, and so made to closely resemble some kinds of genuine caper tea, with which they have been mixed.

Magnetic oxide, sometimes referred to in the trade as iron filings, is found associated with quartz and sand in tea.

An attempt at explaining the presence of this compound has been made by attributing it to the ferruginous character of the soil on which, in some localities, the tea plant has been cultivated, but the quantity found is often much too large to be accounted for by the accidental admixture of portions of the soil during the process of gathering the leaves.

The object of adding Prussian blue, China clay, and similar substances intended to impart a greenish tint appears to be twofold, namely, to improve and give uniformity to the colour of *bonâ-fide* green tea, and also—which is very reprehensible—to give to old and inferior black tea the appearance and external character of green.

In adding the colour, the leaves are usually moistened with rice-water and re-heated. Whilst in this condition, the finely-powdered colour is shaken on the tea, which is continuously stirred, until a uniform tint is obtained. Sometimes a little yellow colour such as turmeric is added first, and then the Prussian blue or indigo; in other instances, the Prussian blue mixture alone appears to have been used.

Some years ago large quantities of exhausted leaves were collected in this country, and, after the addition of gum and other matters, were rolled and re-dried so as to resemble genuine tea. This spurious tea was mixed with tea as imported and sold to the public as genuine, whereby a very serious fraud was perpetrated. By the vigilance of the authorities this practice was suppressed, and we believe that it has never since been successfully resumed. It has occasionally happened that an importation of tea has been submerged in sea-water; in some instances such tea has been re-dried and brought into the market, but perhaps more frequently re-shipped to another country. From the present general soundness, however, of the tea trade, the sale of such a description of tea has been reduced to a minimum.

It is well known that the manufacture of spurious tea has been carried on to a considerable extent in China, and that large quantities of the sophisticated article have been imported from time to time into this country, some of which has been known in the market by the name of the "Mahloo Mixture." Owing, however, to the rigid scrutiny to which tea is now subjected, it is likely that the quantity of such spurious teas passed into this country will become very small.

It has been seen that the adulterants of tea may be conveniently studied under three heads—viz.:

1st. Substances which can be detected by their physical properties.

2nd. Mineral salts, along with some organic substances, all of which have distinctive chemical reactions.

3rd. Partially exhausted leaves.

When tea is treated with hot water a large proportion of the soluble organic matter of the leaf is extracted, together with some of the mineral compounds, chiefly potash salts. If in a suspected sample the problem to be solved was merely whether it consisted wholly of exhausted leaves, the proof of sophistication would be very easy. But when a portion only of the tea is in that condition the question presents greater difficulty. This arises from the fact that in vegetable products, tea amongst the rest, nature has provided no hard-and-fast line as to the proportion in which the soluble or characteristic constituents may be present. Where a given test, say that of the total water extract is relied upon and applied to a number of teas, great difference exists between the maximum and minimum results. It is obvious, therefore, that a considerable admixture of a spurious tea could be added to one of the better kinds without reducing the amount of extract below that obtained from a genuine though inferior tea.

It will be found difficult in practice to separate individual exhausted leaves. Where this can be done, the results of an analysis of them would no doubt be conclusive. Usually it is necessary to weigh out a given quantity—say 100 grains—of the tea, and after repeated extraction with boiling water, the filtered extract is evaporated down either in whole or in part, and the percentage of dry extract calculated.

Or, 200 grains of tea are finely powdered in a mortar, and introduced into a flask with 2,000 grains of water at 60° F. (15·5° C.), and raised to the boiling-point over an argand burner. After this point has been reached, it is removed from the lamp, allowed to stand for 2 minutes, and filtered hot to prevent deposition of the tannate of theine. The specific gravity of the filtrate is then taken with a gravity bottle, and the result noted. The best filter for the purpose appears to be one made of flannel, and used double.

SPECIFIC GRAVITY OF TEA INFUSIONS

Made by extracting 1 part of dry Tea with 10 parts of Water at 60° F. (15·5° C.) raised to the boiling-point and quickly filtered.

Description.	Specific Gravity at 60° F.	Description.	Specific Gravity at 60° F.
Shanghai Congou ...	1009·88	Foochow Souchong ...	1011·44
,, ,, ...	1011·73	Canton Congou ...	1009·97
,, Gunpowder ...	1012·77	,, ,, ...	1011·33
,, Young Hyson	1013·67	,, Scented Caper ...	1011·34
,, ,, ,,	1012·30	,, ,, ,, ...	1013·50
,, Imperial ...	1011·05	,, Orange Pekoe ...	1012·23
,, Hyson ...	1012·74	,, ,, ,, ...	1012·98
,, ,, ...	1012·35	,, ,, ,, ...	1013·54
,, Twankay ...	1012·32	,, ,, ,, ...	1014·51
Foochow Congou ...	1010·91	Indian Young Hyson ...	1013·80
,, ,, ...	1011·13	,, Gunpowder Dust	1014·57
,, Orange Pekoe	1012·51	,, Orange Pekoe ...	1014·32
,, ,, ,,	1013·31	,, ,, ,, ...	1012·89
,, Scented Caper	1013·03	,, Souchong ...	1013·59
,, Oolong ...	1013·83	,, Congou ...	1012·68
,, Souchong ...	1010·97	,, Pekoe Souchong	1012·67

SPECIFIC GRAVITY OF INFUSIONS,
Made by extracting 1 part of dry exhausted Tea with 10 parts of Water at 60° F. (15·5° C.), raised to the boiling-point and quickly filtered.

Description.	Specific Gravity at 60° F.	Description.	Specific Gravity at 60° F.
Congou	1002·42	Hyson	1004·06
Moning	1002·30	Souchong	1002·60
Orange Pekoe	1004·48	"Mahloo Mixture" * ...	1005·72

PERCENTAGE OF EXTRACTIVE MATTER IN TEA DRIED AT 212° F. (100° C.).

Description: Black.	Per Cent. of Extractive.	Description: Green.	Per Cent. of Extractive.
Congou	35·81	Gunpowder	49·03
,, ...	30·94	,, ...	44·54
,, ...	32·59	,, ...	45·55
,, ...	34·64		
,, ...	39·18	Young Hyson	47·10

It will be seen that the range of specific gravity is very considerable in these teas. For purposes of comparison it would, however, be proper to compare the different classes of tea with each other. In the case of Congou (China), the gravities fall between 1009·80 and 1011·70. In green tea and Pekoes they fall between 1011·00 and 1014·50. If, therefore, exhausted leaves were added to good tea, the specific gravity obtained from the mixture would not fall below that given by a genuine poor tea, unless such leaves bore a somewhat high proportion to the genuine tea.

This remark also applies, as has been already pointed out, to the estimation of the total extractive matter in tea, which varies to an equal if not greater extent.

* As imported.

TABLE OF TOTAL AMOUNT OF ASH

In various kinds of Teas dried at 212° F. (100° C.), together with the Proportion of the Ash soluble in Water and insoluble in Acids, &c.

Description.	Moisture.	Ash.	Soluble in Water.	Soluble in Dilute Acids.	Insoluble in Acids, Sand.	Alkalinity as KHO.
Shanghai Congou, common	9·07	7·63	3·23	3·02	1·38	1·67
,, ,, ,,	9·22	5·95	3·55	2·12	·28	1·85
,, ,, fine	8·51	7·41	4·02	2·54	·85	2·10
,, ,, ,,	8·56	6·62	3·72	2·25	·65	1·79
,, Moyune Young Hyson, com...	7·49	7·49	3·83	2·75	·91	2·03
,, ,, ,, ,, fine...	6·90	7·47	3·82	2·52	1·13	1·92
,, ,, Hyson, common	7·29	6·46	3·18	2·43	·85	1·80
,, ,, ,, ,,	7·25	6·55	3·82	2·37	·36	2·03
,, ,, ,, fine	8·65	6·85	3·58	2·29	·98	1·82
Foochow Congou Siftings	8·51	6·87	3·60	2·81	·46	1·78
,, ,, common	9·35	7·34	3·37	2·97	1·00	1·85
,, ,, fine	9·09	6·89	4·01	2·32	·56	2·11
,, Scented Orange Pekoe, com...	7·55	7·17	3·56	2·86	·75	2·01
,, ,, ,, ,, fine...	7·70	6·72	3·99	2·27	·46	1·90
,, Souchong, common	8·45	6·43	2·97	2·76	·70	1·80
,, ,, ,,	9·34	6·18	3·77	2·09	·32	1·90
,, ,, finest	9·36	5·85	3·61	1·95	·29	2·04
,, Moning, common	8·92	6·88	3·55	2·57	·76	1·72
Canton Scented Caper, common	6·87	6·38	3·21	2·49	·68	1·79
,, ,, ,, fine	6·63	6·63	3·47	2·23	·93	1·74
,, ,, Orange Pekoe, com...	7·02	6·63	3·72	2·33	·58	1·99
,, ,, ,, ,, fine...	7·73	7·09	4·10	2·46	·53	2·28
,, Gunpowder	7·08	6·70	3·54	2·54	·62	1·56
Indian, Kaugra Valley Young Hyson	7·36	6·12	3·44	2·25	·43	1·84
,, Pekoe Souchong	8·28	6·28	3·24	2·42	·62	1·90
Himalaya Orange Pekoe	7·84	7·62	3·45	2·83	1·34	1·94
Chittagong ,, ,,	8·92	6·20	4·23	1·77	·20	2·21
Assam ,, ,,	9·41	5·93	3·78	1·93	·22	1·94
,, Pekoe Dust	9·57	6·77	3·55	2·69	·53	2·04
,, ,,	9·35	6·14	3·70	2·15	·29	1·99
Darjeeling ,,	9·06	5·73	3·68	1·86	·19	1·99
Chittagong Congou	9·68	5·66	3·53	1·87	·26	2·27

In the preceding results, which are selected from numerous analyses of average samples of genuine imported teas, the percentage of ash in no case, except one on page 10, reaches 8 per cent., while the sand in only a few instances exceeds 1 per cent.

The amount soluble in water in only one case falls below 3 per cent.; in the majority of the teas the proportion lies between 3·20 and 3·60 per cent. As exhausted tea is deprived of a large part of the soluble salts, it follows that any serious admixture of such leaves would materially lower the percentage of soluble ash.

The following table exhibits the percentage of ash, with the amount soluble in water, etc., found in tea-leaves which had been infused in the ordinary way for domestic use, and afterwards dried:

Percentage of Ash

In used Tea-leaves with the Amount soluble in Water, etc.

Description.	Percentage of Ash.	Soluble in Water.	Soluble in Dilute Acids.	Insoluble in Acids, Sand.	Alkalinity, as KHO.
Congou	3·92	0·54	2·97	0·41	0·13
Moning	4·53	0·85	2·73	0·95	0·33
Orange Pekoe	3·77	0·68	2·52	0·57	0·22
Hyson	5·56	0·76	3·40	1·40	0·25
Souchong	4·12	0·81	2·61	0·70	0·23
"Mahloo Mixture," as imported ...	9·97	1·54	4·12	4·31	0·20

The table on the following page is founded on the analysis of unprepared tea-leaves received direct from India.

TEA.

LEAVES OF THE TEA PLANT, UNPREPARED.

Amount of Extractive, Specific Gravity in the proportion of 1 part of Tea to 10 parts of Water, Ash, etc., calculated on the leaves dried at 212° F. (100° C.).

	Specific Gravity of Infusion of 1 in 10.	Per Cent. of Moisture.	Per Cent. of Extractive.	Per Cent. of Ash.	Percentage of Ash, etc. Soluble in Water.	Soluble in Dilute Acid.	Silica, etc.	Alkalinity (KHO).
Assam Plant:								
Pekoe leaves ...	1015·08	8·32	54·0	5·87	3·49	2·29	·09	1·95
,, Souchong ...	1015·71	8·29	52·3	6·19	3·42	2·66	·11	2·22
,, coarse leaves, larger than used in manufacture ...	1011·87	6·84	50·9	8·50	2·63	5·45	·42	1·30
China Plant:								
Pekoe leaves ...	1013·32	8·84	48·9	5·45	3·36	2·00	·09	1·82
,, Souchong ...	1014·85	8·51	49·3	6·16	3·39	2·63	·14	2·17
,, coarse leaves, larger than used in manufacture ...	1011·96	6·85	39·9	7·57	2·75	4·63	·19	1·36

The Pekoe leaves in the above table show a remarkably high extractive and specific gravity. The coarse leaves, which are larger than those used in the manufacture of tea, give a high total ash, but a low amount soluble in water.

Closely allied to the estimation of the extract of the tea is the determination of the amount of ash soluble in water. When tea is exhausted with water, it is found that the ash is much less rich in potash salts, which, from their greater solubility, are mostly removed by the extraction. The amount of soluble ash in genuine teas ranges from about 2·8 to 4·2 per cent. The presence, therefore, of a low percentage of soluble ash, combined with a low specific gravity, or amount of extract, indicates that exhausted leaves have been added, or that the tea is partly made up of very old leaves.

50 or 100 grains of the tea are incinerated and reduced to an ash in a platinum capsule. The ash is treated with boiling water, filtered, and the insoluble part dried, ignited, and weighed. The alkalinity of the soluble part may then be taken with decinormal sulphuric acid solution, and expressed as potash (KHO).

The amount of mineral matter insoluble in dilute acid (HCl) represents the sand, quartz, etc. The presence of iron and magnetic oxide is indicated by passing a magnet through the powdered sample. If either of these be present it will show itself by gathering round the poles of the magnet. The adhering particles may be analysed for the amount of iron.

If the insoluble ash exceeds much more than 1 per cent., there is evidence of the addition of either sand, quartz, or other earthy impurities.

The substances used for facing tea are best removed by shaking up with cold water a portion of the sample to which facing has been applied, and pouring off the liquid into another vessel before the detached particles have had time to subside. After the insoluble portion has settled it may be collected, dried, and weighed, and the result will afford a means of forming an estimate of the amount of facing present. Care, however, should be taken to make allowance for any sand unavoidably decanted with the colouring matter.

Before applying any chemical test it is well to examine the sediment under the microscope, when Prussian blue, indigo, starch, turmeric, or other added matter will be indicated by their several characteristics.

The colour of the *Prussian blue* disappears under the influence of a little solution of potash, and is restored again on the addition of a little acid.

Indigo, in the presence of potash, remains permanent for some time in the cold, but its colour is discharged by solution of permanganate of potash.

Turmeric is altered to a peculiar reddish brown by alkalis.

Soapstone, which is a silicate of magnesia, requires to be decomposed by fusion with alkaline carbonate; the resulting product is examined for silica and magnesia in the usual way.

China Clay.—If, after fusion, alumina were found, and no magnesia, the presence of China clay would thereby be indicated.

MICROSCOPIC EXAMINATION.

The admixture of foreign leaves with tea is most readily and satisfactorily detected by the microscope. Although it may not be possible to identify every leaf employed to adulterate tea, yet a person intimately acquainted with the structure and characteristics of the tea-leaf will find but little difficulty in distinguishing any foreign leaf which may be present in a sample.

A portion of the tea to be examined should be soaked in hot water until the leaves become soft. The water should then be poured off, and the leaves and fragments of leaves removed to a porcelain slab or piece of glass, carefully opened out, and examined both by the naked eye and also with the aid of a pocket-lens. In this examination attention should be directed to the form of the leaves, the character of the serratures of their margins, and the number, size, and direction of the veins, which often present peculiarities in the leaves of different plants.

The tissues of the leaves are next examined by the microscope, particularly the skin of each surface, as to the form and size of the cells, the presence or absence of hairs, and, if present, their form and structure, also as to the number and form of the stomates on the upper and under surfaces.

The presence or absence of sphæraphides or other plant-crystals should also be noted. In some instances, after the leaf has been softened with boiling water, sufficient of the skin may be obtained for examination by tearing a portion away from an incision made in the leaf with a sharp knife. The best method, however, is to place the leaf in water containing a few drops of nitric acid, and gradually raise the temperature to the boiling-point. The skin then rises in blisters, and may be easily removed by means of a camel's-hair brush.

It is impossible to describe the form and structure of every leaf which might be used as an adulterant of tea; but it may be useful to give a few examples to indicate the way in which different leaves are recognised and distinguished from each other by an examination of their tissues. We have selected for this purpose three leaves—viz. those of the elder, the willow, and the sloe—all of which, as already stated, were at one time used in this country for the sophistication of tea.

34   TEA.

Elder.—The elder-leaf is ovate in form and pointed at the apex, with one lobe somewhat larger than the other at the base.

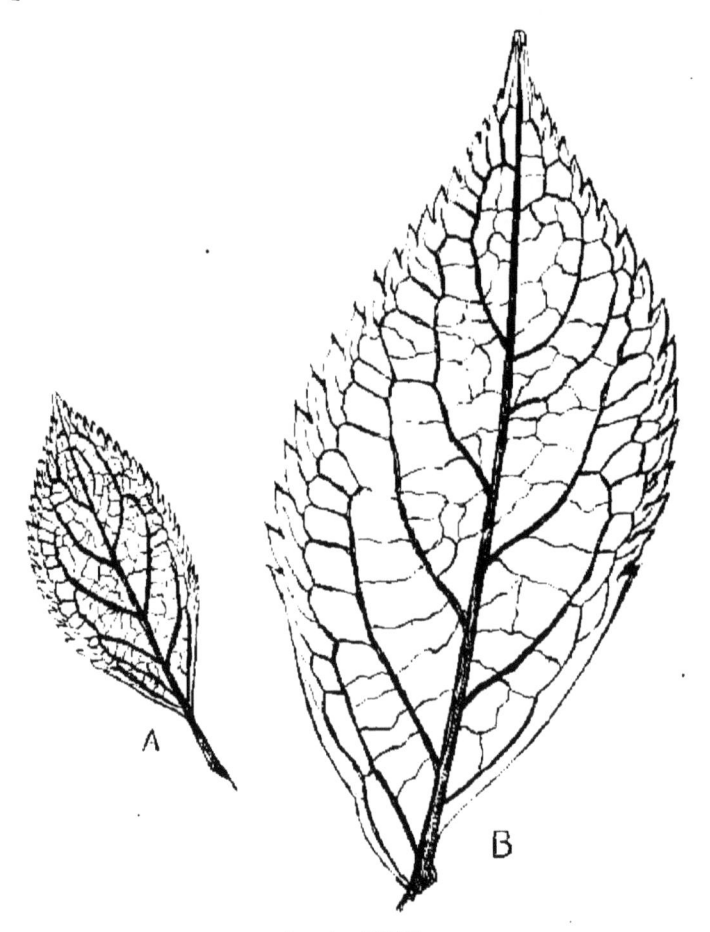

FIG. 5.—ELDER.

Its margin is regularly and sharply toothed, except at the base of the leaf (Fig. 5)—A, medium-sized leaf; B, the same magnified.

The cells of the skin are not so sinuous as those of tea, and are

TEA. 35

distinctly striated. There are two kinds of very characteristic hairs found on the elder-leaf, one short and conical, and the other jointed and bulbous. These hairs are more numerous on the upper than on the under surface, and cannot in any case be mistaken for those of tea (Fig. 6).

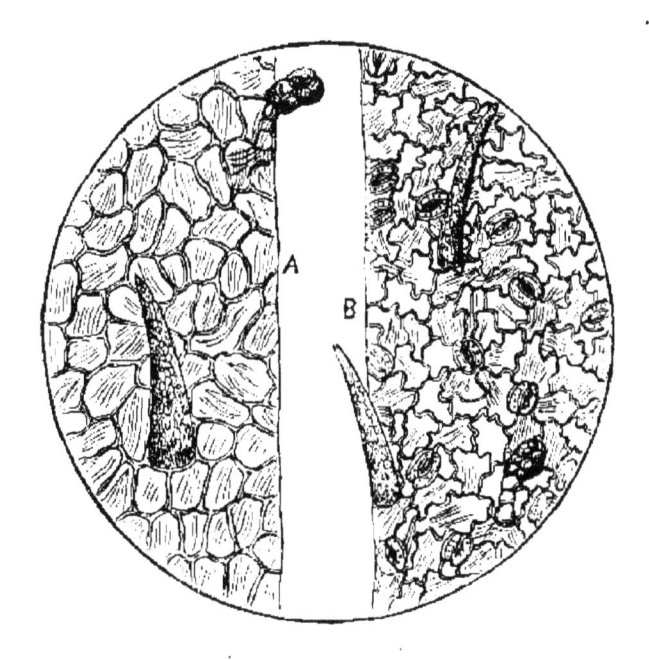

FIG. 6.—SKIN OF THE UPPER (A) AND UNDER (B) SURFACES OF ELDER-LEAF.

36 TEA.

Willow.—The leaves of the common white willow are elliptical-lanceolate in form and acute at the apex (Fig. 7)—A, medium-sized leaf; B, the same magnified. The margin of the leaf is

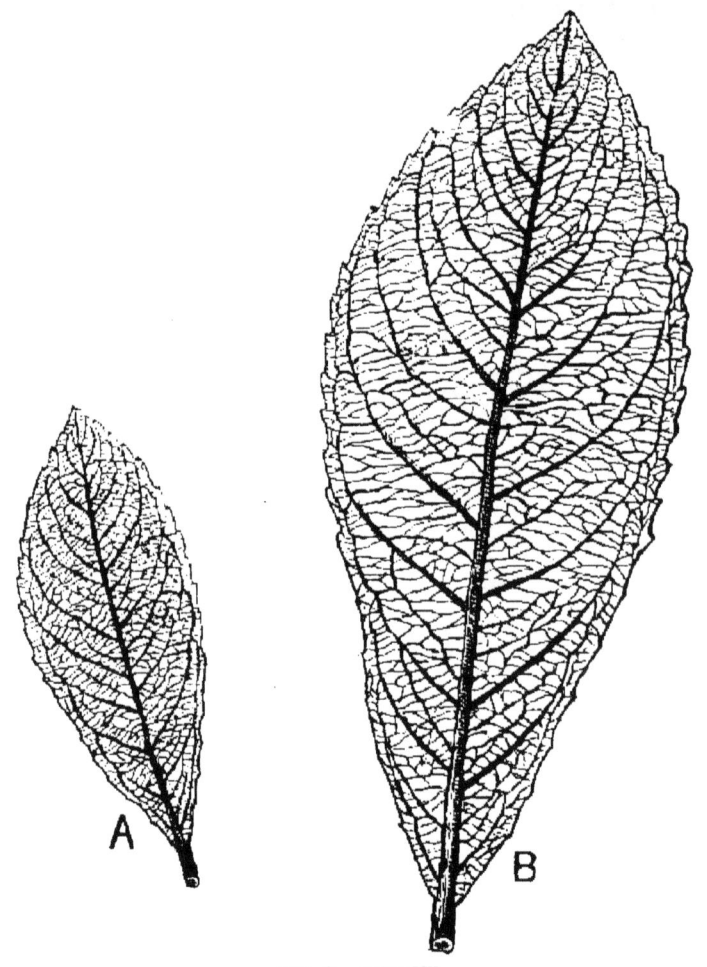

FIG. 7.—WILLOW.

serrated, the lower serratures being somewhat glandular. The cells of the epidermis are much smaller than in tea, and are

TEA. 37

not sinuous. The hairs, which are so abundant on both sides of the leaf as to give it a silky appearance, are unicellular, coarse, and very tortuous. They cannot be mistaken for those of tea (Fig. 8).

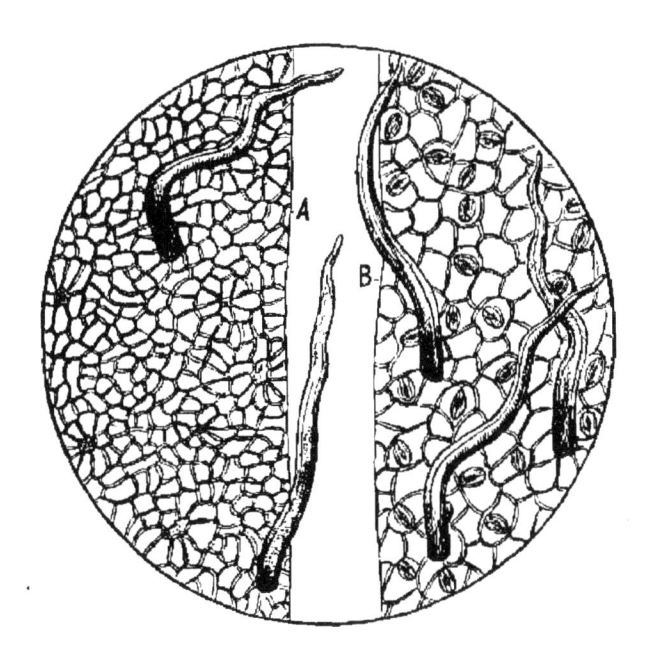

FIG. 8.—SKIN OF THE UPPER (A) AND UNDER (B) SURFACES OF WILLOW-LEAF.

38   *TEA.*

Sloe.—The leaf of the sloe somewhat resembles that of the tea in size and form. Its margin, however, is distinctly serrated,

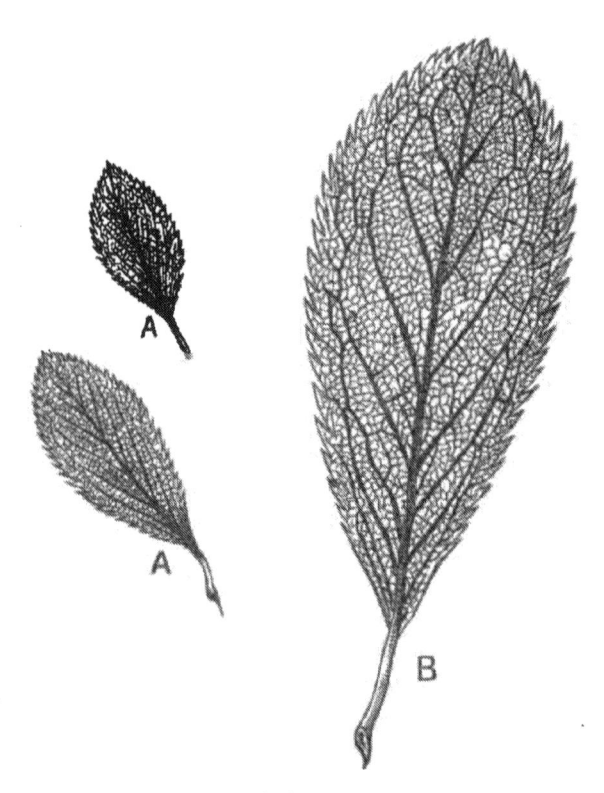

FIG. 9.—SLOE.

the serratures being very numerous, coarse, and irregular (Fig. 9) —A, medium-sized leaves; B, leaf magnified.

The cells of the epidermis are not sinuous like those of tea, and are much smaller, especially on the under surface of the leaf. The

TEA. 39

stomates are also smaller and less numerous (Fig. 10, B). The cells on the upper surface are striated (Fig. 10, A).

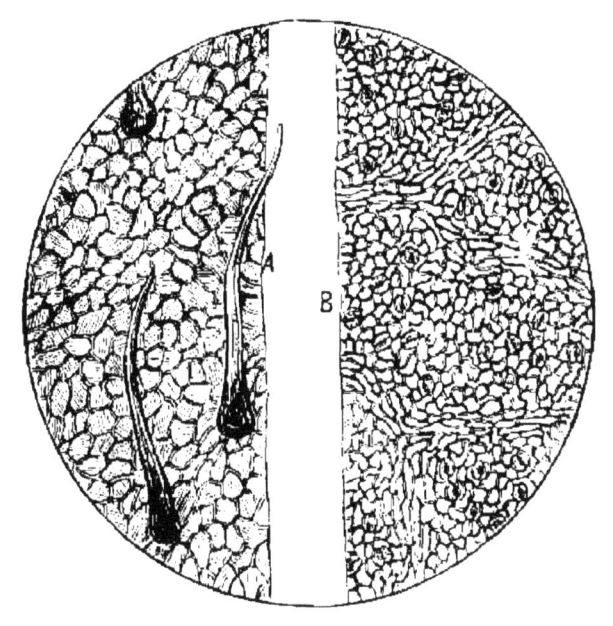

FIG. 10.—SKIN OF THE UPPER (A) AND UNDER (B) SURFACES OF SLOE-LEAF.

Hairs are found in great abundance on the midrib, the veins, and the margin of the leaf. They are shorter and stouter than those of the tea-leaf, and are somewhat club-shaped at the base.

The serrated edge of the leaf—the striated cells of the upper surface, and the short stout hairs are sufficiently characteristic to distinguish the sloe from the tea leaf.

COFFEE.

BOTANICAL ORIGIN.—The coffee tree, *Caffea arabica*, belongs to the natural order *Cinchonaceæ*. It is an evergreen, with smooth, shining, oblong and leathery leaves. It was originally a native of Abyssinia and Arabia, but has been naturalised in most of the tropical countries colonised by Europeans. In its natural state it attains to a height ranging from 15 to 20 feet, but in cultivation it is pruned so as to remain about 6 feet high to facilitate the gathering of the berries. The coffee tree continues flowering for eight months in the year, and during the whole of that period fruit of very unequal ripeness is found upon its branches. Three gatherings of the berries are made in the course of the year.

Description.—The ripe berry is about the size and shape of a small cherry, and of a dark scarlet colour. In each berry there are usually two beans placed face to face, and enclosed in a hard coriaceous membrane, surrounded, when fresh, by a fleshy shell or pericarp, which, when dry, becomes hard and brittle. The berries, after being dried by the rays of the sun, are passed between rollers to remove the dried pulpy matter and the coriaceous membrane. They are then sorted according to size preparatory to roasting. The tissues enclosing the bean are generally distinguished as the husk, the parchment, and the skin. The first two are very rarely if ever met with in commerce, but the skin, which is more or less entangled in the folds of the longitudinal furrow on the flat side of the bean, is present in every sample of commercial coffee.

History.—It is affirmed that the use of coffee has been common in Abyssinia from time immemorial; but the early days of coffee drinking are involved in considerable obscurity. The employment of coffee as a beverage was first commenced in England about the middle of the seventeenth century, when a coffee-shop was opened in the city of London, and similar establishments gradually sprang up, notwithstanding the heavy tax placed on coffee, and the disfavour with which they were for a long time looked upon by the constituted authorities.

During the year 1699, the consumption of coffee in the United Kingdom amounted to 100 tons, of which 70 tons were consumed in England. In the year 1808, the consumption amounted to 477 tons, and a reduction of duty in that year was followed by an immediate increase in consumption, from 1,069,696 lbs. to 9,251,837 lbs. In 1825 another reduction of the duty took place, after which the consumption still further increased, until, in 1847, the maximum of 37,441,373 lbs. was reached. Since that date the use of coffee has greatly declined, and in 1874 the consumption was 31,859,408 lbs., whilst in 1880 it amounted to 32,480,000 lbs. The falling off which has taken place since 1847 is supposed to be due partly to an extended use of tea, and partly to the introduction of coffee substitutes.

The cultivation of the coffee plant began to extend towards the end of the seventeenth century, and soon afterwards its growth was successfully carried on in various countries possessing tropical climates, such as Java, Ceylon, Jamaica, and Brazil. The cultivation of coffee in Brazil is now very considerable, but that produced in Java and Ceylon is of superior quality, though the Mocha coffee is the finest of all. The quantity of "Mocha" which finds its way into Europe from Arabia is said to be comparatively small, and we are now indebted chiefly to India for our supply of this description of coffee.

The effect of coffee on the human system is to counteract the tendency to sleep; and it is almost certain that it was this property which originally led to its use as a beverage. It also excites the nervous system, and when taken in excess produces contractions and tremors of the muscles, and a feeling of buoyancy and exhilaration somewhat similar to that produced by alcohol, but does not end with depression or collapse.

CHEMICAL COMPOSITION.

The principal organic substances composing the raw coffee beans are caffeine, fat, caffeic acid, gum, saccharine matter, legumin, and cellulose. These substances are also found in the roasted beans, though modified in some degree by the roasting process. By torrefaction the woody tissue of the fresh bean, which is hard and horny, undergoes a considerable change, and becomes friable, and the difficulty of pulverising and exhausting it by water is greatly diminished.

Coffee has, like tea, been the subject of numerous investigations, but the following analysis by Payen is most frequently quoted:

Cellular Tissue	34·00
Hygroscopic Moisture	12·00
Fat	13·00
Starch, Sugar, Dextrin, and Vegetable Acids	15·50
Legumin	10·00
Chlorogenate of Potash and Caffeine	3·5 to 5·0
Nitrogenous portion	3·00
Free Caffeine	0·80
Thick insoluble Ethereal Oil	0·001
Aromatic Oil	0·002
Mineral Constituents	6·697

We have obtained the following results from two samples of coffee, both in the raw and roasted condition:

	Mocha.		East Indian.	
	Raw.	Roasted.	Raw.	Roasted.
Caffeine	1·08	·82	1·11	1·05
Saccharine Matter	9·55	·43	8·90	·41
Caffeic Acids	8·46	4·74	9·58	4·52
Alcohol Extract, containing Nitrogenous and Colouring Matter	6·90	14·14	4·31	12·67
Fat and Oil	12·60	13·59	11·81	13·41
Legumin or Albumin	9·87	11·23	11·23	13·13
Dextrin	·87	1·24	·84	1·38
Cellulose and Insoluble Colouring Matter	37·95	48·62	38·60	47·42
Ash	3·74	4·56	3·98	4·88
Moisture	8·98	0·63	9·64	1·13
	100·00	100·00	100·00	100·00

Caffeine, $C_8 H_{10} N_4 O_2$.—This, from its dietetic properties, is one of the most valuable constituents of coffee, and is identical with the alkaloid found in tea, and described under the name of theine. Caffeine was discovered in coffee by Runge, in the year 1820; by Oudry, in tea in 1827; by Martin, in guarana, in 1840; and by Stenhouse, in *Paullinia*, in 1843. Its chemical constitution was settled by Jobst and Mulder.

Fat.—The fatty portion of the coffee consists of a mixture of several fats, some of which have the consistence of an oil. These become altered to some extent in the process of roasting, and a portion of the volatile fatty acid is set free, and changed in character by the heat. The quantity of fat in the raw coffee is also more or less affected by the roasting, but the extent to which this takes place depends upon the degree to which the torrefaction is carried. We have found that in ordinary commercial samples of roasted coffee the loss of fat amounts to about 1 per cent., but the loss may be greater than this in cases in which the coffee is subjected to a more than ordinary degree of roasting.

Caffeone.—This remarkable and characteristic body is a product of the roasting of coffee. It consists of an oil which gives the coffee aroma, and which is doubtless a mixture of the altered volatile oil and other products probably obtained from several of the constituents of the raw coffee by the action of heat. Caffeone, which is the aromatic principle of coffee, can be partially separated by distillation, when it is found as a brown oil heavier than water.

Caffeic Acid.—This is also known by the names of caffeotannic acid and chlorogenic acid, and is present in coffee to the extent of from 3 to 5 per cent. It is obtained as a yellowish mass, and is more easily soluble in water than in alcohol. When heated it emits the peculiar odour of coffee, and from this it is thought that this acid plays an important part in developing the flavour found in the roasted beans. An infusion of raw coffee, when heated with alkalis, gradually acquires a bright green colour, which is supposed to be produced by the change of caffeic acid into viridic acid. Caffeic acid, distilled with peroxide of manganese and sulphuric acid, yields quinone.

Sugar.—The raw coffee beans contain from 5·70 to 7·70 per cent. of a fermentable sugar. When, however, the sugar extracted from the beans is boiled for a few minutes with a little dilute sulphuric acid, the amount shown by an alkaline cupric solution is equivalent to from 8·9 to 9·55 per cent.

The sugar in coffee, unlike cane-sugar in sweet roots, becomes almost wholly converted into caramel in the process of roasting; but, so far as we are aware, no satisfactory explanation has been given of this peculiarity.

It has been suggested that a portion of the sugar exists in the form of a conjugate combination, like the sugar in salicin and tannin; but we have failed to find any confirmation of this view from the results of our experiments. On the contrary, our results point to the presence of a sugar peculiar to coffee, and bearing somewhat the same relation to cane sugar as melezitose, mycose, and similar compounds. When the saccharine extract of the beans is boiled with a little dilute sulphuric acid, a quantity of glucose is obtained nearly equivalent to the weight of dry extract operated upon—a result which is inconsistent with the existence of sugar as a glucoside. The sugar is also but slightly affected by boiling even for a considerable time with acetic acid. If it existed as cane-sugar it would, under such conditions, be converted into invert sugar.

Gum.—This appears to be present in coffee in the form of dextrin, and can be recovered and identified in the water extract after the coffee has been thoroughly exhausted with ether and alcohol. The amount of gum is but little altered in the roasting. The quantity present is from ·84 per cent. in the raw to 1·38 per cent. in the roasted berries.

Albumin.—The amount of soluble albumin or legumin is greater in coffee than in tea. The amount found is 2·53 per cent. in the raw and 1·47 per cent. in the roasted coffee.

Cellulose.—The cellulose of coffee berries is of a very hard and horny character, and said by some not to be converted into sugar by boiling with sulphuric acid. We found, however, that by treating the cellulose with a sulphuric acid solution of the strength of 1 in 2, a quantity of sugar was obtained, as indicated by a standard copper solution, equivalent to nearly 36 per cent. of the quantity of coffee analysed.

COFFEE.

Ash.—The following table, which is taken from the Coffee Report by Messrs. Graham, Stenhouse, and Campbell, exhibits the composition of the ash of seven kinds of coffee:

ANALYSIS OF THE ASH OF COFFEE.

	Plantation Ceylon.	Native Ceylon.	Java.	Costa Rica.	Jamaica.	Mocha.	Neilgherry.
Potash	55·10	52·72	54·00	53·20	53·72	51·52	55·80
Soda	—	—	—	—	—	—	—
Lime	4·10	4·58	4·11	4·61	6·16	5·87	5·68
Magnesia	8·42	8·46	8·20	8·66	8·37	8·87	8·49
Sesquioxide of Iron	0·45	0·98	0·73	0·63	0·44	0·44	0·61
Sulphuric Acid	3·62	4·48	3·49	3·82	3·10	5·26	3·09
Chlorine	1·11	0·45	0·26	1·00	0·72	0·59	0·60
Carbonic Acid	17·47	16·93	18·13	16·34	16·54	16·98	14·92
Phosphoric Acid	10·36	11·60	11·05	10·80	11·13	10·15	10·85
Silica, etc.	—	—	—	—	—	—	—
Sand	—	—	—	—	—	—	—

MICROSCOPIC STRUCTURE.

The structure of the coffee berry is comparatively simple.

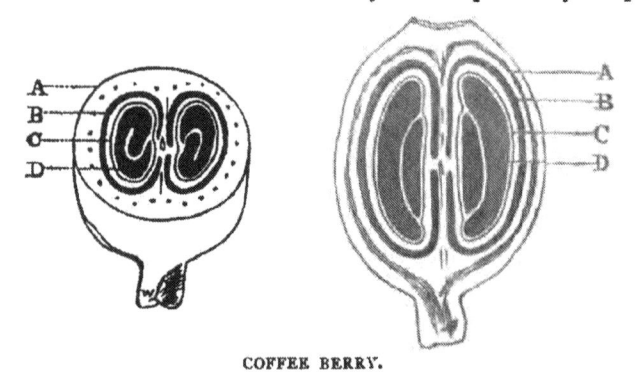

COFFEE BERRY.
FIG. 11.—TRANSVERSE SECTION. FIG. 12.—LONGITUDINAL SECTION.

Transverse and longitudinal sections of it are shown in Figs. 11 and 12. The husk is represented at A, the parchment at B, the skin at C, and the bean at D.

The bean is the only portion that is imported into this

country, the three outer membranes being removed in the preparation of the coffee for market. A portion, however, of the membrane, called the skin, is always more or less entangled in the folds of the bean, and is present in variable proportions in all genuine coffee. We have, therefore, to deal only with the microscopic structure of the skin and bean.

The skin is a thin delicate membrane, lined with peculiar

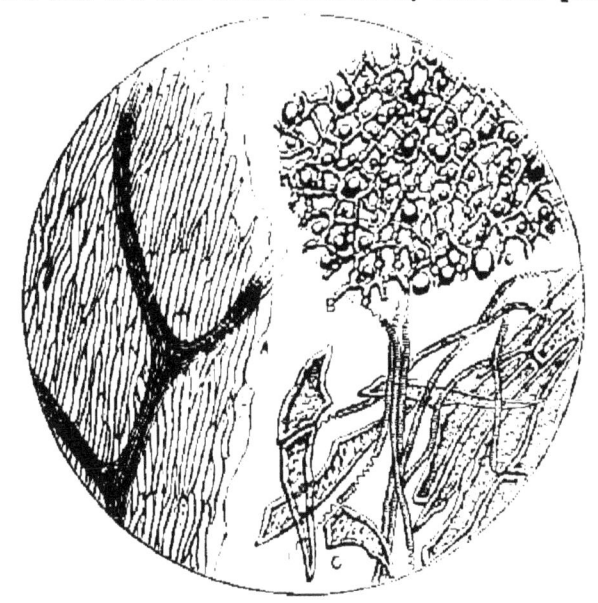

FIG. 13.—COFFEE.

(A) Skin of Bean ... ... magnified 50 diameters.
(B) Section of Bean ... ... ,, 125 ,,
(C) Cells composing Skin of Bean ... ,, 200 ,,

spindle-shaped bodies, lying in rather a confused manner, and forming a kind of plate, as shown at A in Fig. 13. These cells are very thick-sided, and have a central canal, their edges presenting the appearance of being marked by minute transverse bars. When more highly magnified, the bars are seen to resemble open ducts, as represented at C.

The structure of the bean is represented at B. It is composed entirely of a mass of quadrangular, hexangular, and irregular cells, with very thick walls or sides enclosing oil globules. These walls are somewhat sinuous, and are characterised by a quantity of irregular projections on their sides, which, when examined in the fresh state, present a luminous appearance, but, when roasted, appear as opaque masses, imparting a rugged character to the cell walls. The projections are due in some cases to very minute starch granules, while in others they appear to belong to the solid sides of the cell walls. These are the only vegetable structures that occur in commercial coffee. The tissues are not altered by the roasting, and, excepting that they are partially charred, still preserve their characteristic structure. The few starch granules which are present in the fresh bean are, however, entirely destroyed.

ANALYSIS.

Fat.—One hundred grains of dry and finely rasped or ground coffee are repeatedly exhausted with ether. The ether is evaporated in a weighed beaker, and the weight of the dry residue ascertained. The non-fatty substances, which may have been also removed by the ether, are dissolved out with boiling water. The aqueous portion is evaporated and weighed, and the amount deducted from the first weight. The difference is the proportion of fat present in the coffee.

Caffeine.—The caffeine is estimated in the same way as that described under "theine" on page 16.

Caffeic or Caffeo-tannic Acid.—One hundred grains of coffee are exhausted with alcohol. The alcohol is evaporated, and to the aqueous residue subacetate of lead is added. The precipitate is thrown on a filter, washed and decomposed with sulphuretted hydrogen. The filtrate from the plumbic sulphide is evaporated to dryness, when the caffeic acid is obtained as a yellowish brittle mass.

A larger amount of astringent matter is obtained by extracting, with ether in presence of phosphoric acid, the non-fatty portion of the alcohol extract of the coffee. We have obtained by this means from 8·4 to 9·5 per cent. in the raw coffee.

Sugar.—The estimation of the sugar in coffee is sometimes made by the copper test described on pages 56 and 106, but the quantity is more frequently determined by what is termed the fermentation test. In applying the latter method it is usual to take from 1,000 to 2,000 grains of the sample, exhausting at least four times with hot water, and adding the extracts together. To the solution 250 grains of pressed yeast are then added, and the whole set to ferment for 48 hours, at a temperature of from 70° to 80° F. 250 grains of the yeast are next added to a quantity of distilled water equal in bulk to the coffee extract, and fermented under like conditions. The solutions are then distilled, and the specific gravity and bulk of the distillates taken. The amount of alcohol in the yeast solution is deducted from the amount found in the coffee extract, and the sugar calculated from the remainder, as described on page 108, under the head of "Sugar."

Albumin.—The albumin is determined by making a nitrogen combustion of the dry residue of the coffee berries after they have been exhausted with ether, alcohol, and water. The soluble albumin is found with the dextrin and gummy precipitate given by alcohol in the water extract, and is estimated by finding the amount of nitrogen, and multiplying the product by 6·3.

Gum or Dextrin.—The ground coffee is exhausted with ether, then with alcohol, and the residue treated several times with boiling water. The aqueous extract is boiled for some time to precipitate albumin, filtered, and to the filtrate is added 200 grains of concentrated hydrochloric acid. The solution is now treated with alcohol to precipitate the gum. The precipitate is thrown on a tared filter washed with alcohol, and then dried and weighed.

Cellulose.—This is estimated indirectly, as in the case of the cellulose of tea. The coffee is treated successively with ether, alcohol, and water, and the residue is afterwards dried and weighed. From the dry residue thus obtained is deducted the amount of albumin and ash found therein, and the remainder represents the percentage of cellulose in the coffee.

ADULTERATION.

In dealing with the adulteration of coffee, an important consideration meets us which does not apply in the case of tea. For many years the public have shown a preference for a certain substitute for, or rather addition to, coffee which is known under the name of "chicory." This may legally consist of the roasted chicory root itself or of an allied root, or other vegetable substance applicable to the uses of chicory.

The sale, therefore, of a mixture of chicory and coffee in a case where the fact of such mixture is stated cannot be regarded as an adulteration. It is only in cases in which mixtures are sold as coffee that it becomes necessary to consider chicory as properly among the adulterants of coffee.

The substances which have been generally employed for mixing with or adulterating coffee are:

1st. Roots—such as chicory, dandelion, mangold wurzel, turnips, parsnips, and carrots.

2nd. Seeds—such as beans, peas, date-stones, acorns, malt, rye, etc.

3rd. Burnt sugar, biscuits, locust beans, and figs.

Burnt sugar or caramel is sometimes added to coffee to give colour to the extract; but it is now frequently used, it is said, to improve and conserve the aroma of the roasted berry. When employed for this latter purpose a small quantity of raw sugar, not exceeding 3 lbs. per hundredweight, is allowed to be mixed and roasted with the raw coffee berries.

The addition of oxide of iron, or some ferruginous earth, to give colour or weight, has been detected; but at the present time this form of adulteration, if practised at all, is very rare.

The vegetable substance is in all cases roasted and ground to the consistence of coffee; and, in some instances, the article is first used to add to, or adulterate chicory, and in such form mixed with the coffee.

In the chemical composition of the extract of chicory there

is but little in common with that of coffee, and chicory is used rather as an addition to, than as a substitute for coffee. Indeed, it may be said that sugar is almost the only constituent which chicory and the allied roots contain that causes them to have any property in common with coffee, for in nearly all other respects they are different. The quantity of sugar contained in chicory and the like roots is considerable, and in the process of roasting a large proportion of it is converted into caramel, which imparts to their extract the bitter of burnt sugar with a somewhat similar aroma. It is admittedly the taste of this caramel bitter which recommends such an extensive use of chicory and other roots as an addition to coffee.

As yet there has been no seed found which, when roasted and ground, corresponds with coffee either in its physiological properties or in the chemical composition of its extract, and the divergence between coffee and the seeds above enumerated is quite as great as between roots and coffee.

In the detection of the adulteration of coffee, advantage is taken of a difference between both its physical and chemical properties and those of its adulterants.

Thus the large quantity of caramel produced in chicory and some other roots in the process of roasting, furnishes the means of applying a simple and convenient preliminary test for detecting their presence in coffee.

When a few grains of coffee containing chicory are placed on the surface of water in a test-tube or wine-glass, each particle of chicory becomes surrounded by a yellowish-brown-coloured cloud which rapidly diffuses itself in streaks through the water, till the whole acquires a brownish colour.

Other sweet roots when present will produce under like conditions the same effect as chicory, but not so rapidly or perceptibly as the latter. Pure coffee under similar conditions gives no sensible colour to the water until after the lapse of about a quarter of an hour. The relative colouring power of coffee, chicory, and a variety of other vegetable substances used in the adulteration of coffee was determined by Messrs. Graham, Stenhouse, and Campbell. This was done by infusing equal quantities of each substance in water, as in the preparation of coffee for domestic use, filtering the infusions and observing the colour in glass tubes of about 1 inch in diameter. The solutions, which were prepared at 212° F. (100° C.), were made very dilute, and, for comparison, a standard solution was prepared by dissolving 1 part of caramel, carefully made from cane-sugar, in 2,000 parts of water. To produce a depth of colour equal to that of the standard solution, a larger proportion of the adulterating substance is required, than the 1 in 2,000 of the standard.

The following table exhibits the actual weights required of each substance when roasted and prepared in imitation of coffee, to be dissolved in 2,000 parts of water, to produce an equal depth of colour:

Substance	Weight
Caramel	1·00
Mangold Wurzel	1·66
Bouka (a coffee substitute)	1·66
Sparke's Vinegar Colouring	1·74
Black Malt	1·82
White Turnips	2·00
Carrots	2·00
Chicory (darkest Yorkshire)	2·22
Parsnips	2·50
Maize	2·86
Rye	2·86
Dandelion Root	3·33
Red Beet	3·33
Bread Raspings	3·64
Acorns	5·00
Over-roasted Coffee	5·46
Highly-roasted Coffee	5·77
Medium-roasted Coffee	6·95
Another sample of Coffee	6·66
White Lupin Seed	10·00
Peas	13·33
Beans	13·33
Spent Tan	33·00
Brown Malt	40·00

It will thus be seen that 2·00 parts of carrot, or 2·22 of chicory, have the same colouring power as 6·95 parts of medium-roasted, or 5·77 of highly-roasted coffee, or of 3·64 parts of bread raspings, or 13·33 parts of roasted beans.

Another respect in which infusions of coffee, chicory, and some other substances differ from one another is that of their specific gravities. When, in a sample of coffee, the adulterant has been identified by a microscopical examination, the difference in the densities of the infusions becomes immediately available for estimating the proportion present. In preparing the infusion, 100 grains of the ground substance are placed in 1,000 grains by measure of distilled water. The temperature of the mixture is then raised to the boiling-point, where it is maintained for half a minute. The resulting infusion is next filtered, and the specific gravity of the filtrate taken at a temperature of 60° F. (15·5° C.). This will be found to vary according to the substance employed, in the manner shown in the subjoined table:

SPECIFIC GRAVITIES OF SOLUTIONS AT 60° F.
One part of Substance to 10 parts of Water.

Substance	Specific Gravity
Spent Tan	1002·1
Date Stones	1002·9
Lupin Seed	1005·7
Acorns	1007·3
Peas	1007·3
Mocha Coffee	1008·0
Beans	1008·4
Neilgherry Coffee	1008·4
Plantation Ceylon Coffee	1008·7
Java Coffee	1008·7
Jamaica Coffee	1008·7
Costa Rica Coffee	1008·9
Native Ceylon Coffee	1009·0
Costa Rica Coffee	1009·5
Parsnips	1014·3
Carrots	1017·1
Black Malt	1021·2
Turnips	1021·4
Rye Malt	1021·6
English Chicory	1021·7

COFFEE.

SPECIFIC GRAVITIES OF SOLUTIONS AT 60° F.—*Continued.*

Dandelion Root	1021·9
Red Beet	1022·1
Foreign Chicory	1022·6
Guernsey Chicory	1023·2
Mangold Wurzel	1023·5
Roasted Figs	1024·9
Maize	1025·3
Bread Raspings	1026·3
British Gum	1037·9
Gum Arabic	1038·6

It will be seen from this table that the low specific gravity of the coffee infusion distinguishes it from the roots and cereals. If an infusion of a sample of coffee containing 20 per cent. of rye malt were prepared in the manner described, its specific gravity would rise above 1009·5—the number corresponding to the infusion of coffee of the highest density—and would approach 1021·6—the number corresponding to rye malt—almost exactly in the proportion in which the latter was present.

This may be illustrated (with coffee and chicory) as follows:

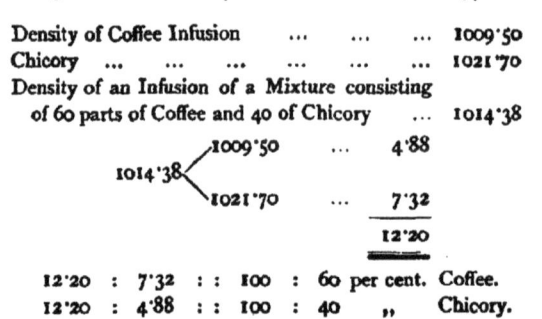

Density of Coffee Infusion	1009·50
Chicory	1021·70
Density of an Infusion of a Mixture consisting of 60 parts of Coffee and 40 of Chicory	1014·38

$$1014\cdot38 \begin{cases} 1009\cdot50 \quad \ldots \quad 4\cdot88 \\ 1021\cdot70 \quad \ldots \quad 7\cdot32 \end{cases}$$
$$\overline{12\cdot20}$$

12·20 : 7·32 : : 100 : 60 per cent. Coffee.
12·20 : 4·88 : : 100 : 40 ,, Chicory.

There is a considerable difference between the quantity of sugar present in roasted coffee and in many of the vegetable substances used for its adulteration, particularly chicory and the other sweet roots, and it is sometimes practicable to make this the means of affording confirmatory proof of adulteration.

The amount of sugar in coffee, and in a great variety of vegetable substances, has been determined, both before and after roasting, by the process of fermentation and distillation.

The results obtained by Messrs. Graham, Stenhouse, and Campbell are given in the following tables:

SUGAR IN COFFEE BEFORE AND AFTER TORREFACTION.

Varieties of Coffee.	Sugar per Cent. Raw.	Sugar per Cent. Roasted.
1. Plantation Ceylon	7·52	1·14
2. ,, ,,	7·48	0·63
3. ,, ,,	7·70	0·0
4. ,, ,,	7·10	0·0
5. Native Ceylon	5·70	0·46
6. Java	6·73	0·48
7. Costa Rica	6·72	0·49
8. ,, ,,	6·87	0·40
9. Jamaica	7·78	0·0
10. Mocha	7·40	0·50
11. ,,	6·40	0·0
12. Neilgherry	6·20	0·0

SUGAR IN CHICORY AND OTHER SWEET ROOTS BEFORE AND AFTER TORREFACTION.

	Sugar per Cent. Raw.	Sugar per Cent. Roasted.
Foreign Chicory	23·76	11·98
Guernsey ,,	30·49	15·96
English ,,	35·23	17·98
English ,, (Yorkshire)	32·06	9·86
Mangold Wurzel	23·68	9·96
Carrots (ordinary)	31·98	11·53
Turnips ,,	30·48	9·65
Beet Root (red)	24·06	17·24
Dandelion Root	21·96	9·08
Parsnips	21·70	6·98

The fermentation test has the advantage of being easily applied, and the results might, in combination with the microscope, prove useful in some cases. In applying the test, the process described on page 49, under the head of "Analysis," is followed.

The determination of the sugar by the copper test is more tedious than by the fermentation test; but at the same time it is not open to some of the objections which may be urged against the other. In estimating the sugar by this method, 200 grains of the coffee are exhausted with hot water: to the extract is added basic acetate of lead, so long as a precipitate appears. The extract is then filtered and the precipitate washed with hot water. The filtrate is next treated with sulphuretted hydrogen, and again filtered and the lead precipitate washed. The solution is now boiled to expel sulphuretted hydrogen, and tested for glucose with a standard solution of copper, as described on page 106.

There are some cases in which the character of the ash affords evidence of adulteration. The ash of coffee is distinguished by the absence of soda, and also by the extremely small proportion of silica it contains, the amount of which varies from 0·17 to 0·45 per cent. On the other hand, the amount of silica or sand in the ash of chicory and dandelion varies from 10·69 to 35·85 per cent.; barley, from 17·3 to 32·7; oats, from 37·82 to 50·28; and rye, from 0·69 to 14·6. Thus it is sometimes practicable to obtain valuable information without making a formal analysis of the ash, by simply digesting it in hydrochloric acid, and observing the character and quantity of what remains insoluble. The presence of 1 per cent. or upwards of silica in the ash of a sample is sufficient to raise a suspicion of adulteration. It is not often that an analysis of the ash of a suspected sample is resorted to in practice; it is a tedious operation, and would only be performed with the view of confirming the results obtained by some of the other methods of examination to which reference has been made.

The following table, which is taken from the Coffee Report by Messrs. Graham, Stenhouse, and Campbell, exhibits a complete analysis of the ash of four kinds of chicory, also of lupins, acorns, maize, parsnips, and dandelion root.

ANALYSES OF CHICORY AND CERTAIN SEEDS AND ROOTS.*

	Chicory, deducting Sand and Silica.				Deducting Silica, etc.					Not deducting Silica, etc.				
	Darkest English Yorkshire.	English.	Foreign.	Guernsey.	Lupins.	Acorns.	Maize.	Parsnips.	Dandelion Root.	Lupins.	Acorns.	Maize.	Parsnips.	Dandelion Root.
Potash	38.53	27.85	46.07	46.27	33.83	55.49	31.28	56.86	20.22	33.54	54.93	30.74	56.54	17.95
Soda	9.34	16.90	3.17	5.49	17.90	0.63	—	—	34.87	17.75	0.63	—	—	30.95
Lime	10.79	10.81	7.78	7.65	7.81	6.98	3.11	6.88	12.87	7.75	6.01	3.06	6.85	11.43
Magnesia	6.06	8.08	5.33	5.55	6.23	4.36	14.98	6.52	1.47	6.18	4.32	14.72	6.49	1.31
Sesquioxide of Iron	4.38	3.50	8.29	5.08	—	0.54	0.85	0.53	1.42	—	0.54	0.84	0.53	1.27
Sulphuric Acid	11.38	11.78	8.38	8.67	6.85	4.83	4.20	4.09	2.66	6.80	4.79	4.13	4.07	2.37
Chlorine	5.67	5.23	5.03	6.58	2.12	2.53	0.50	2.10	4.32	2.11	2.51	0.50	2.09	3.84
Carbonic Acid	2.04	3.22	4.36	4.60	0.56	13.82	—	11.50	6.99	0.56	13.69	—	11.44	6.21
Phosphoric Acid	12.27	12.61	11.00	9.59	25.74	11.26	45.29	13.91	12.63	25.53	11.15	44.50	13.84	11.21
Silica, etc.	3.81	2.61	12.75	10.52	—	—	—	—	—	0.87	1.01	1.78	0.57	11.26
Sand	9.32	8.08	23.10	20.19	—	—	—	—	—	—	—	—	—	—

The relative solubility in ether of coffee and a variety of vegetable substances used to adulterate it, may, in some cases, be found useful as a confirmatory test for indicating the character and the proportion in which the adulterating substance is present. Coffee yields much more soluble matter to ether than do sweet roots, cereals, and leguminous seeds generally. Thus roasted coffee, when agitated four times successively in ten times its weight

* These Analyses, and also those on p. 46, were made by the Author for the Coffee Report.

of ether, gives from 14·79 to 15·10 per cent. of oil and resin, including a small proportion of caffeine. Roasted chicory, when similarly treated, gives 7·72 per cent. of extract. Roasted maize gives 4·30 per cent., and roasted beans 1·57 per cent. The deficiency of fat in the sweet roots and other vegetable substances is frequently made up in the roasting by the addition of American or Australian tallow to prevent burning, and in applying the ether test it is therefore necessary to guard against being misled by the presence of a small proportion of foreign fat.

Venetian Red or Oxide of Iron.—It is very rare, as already stated, that any mineral matter such as oxide of iron is now employed to give colour or weight to coffee. The oxide is usually in such a fine state of division that it cannot be seen by the naked eye, or be detected and identified by the microscope. When the presence of such a compound is suspected it is necessary to burn a given weight of the coffee and observe whether the ash is of a deep red or yellowish-brown colour, and if so it is probable that oxide of iron has been added, and it will be necessary to determine the amount of iron in the ash and compare the result with the quantity found in genuine coffee, or the proportion that may be present in mixtures of chicory and coffee.

Burnt Sugar.—When a sample of coffee which is free from roots and other foreign vegetable substances, imparts to water a deep and rapid colouration, there is reason to suspect that burnt sugar has been added. The black shining particles of the caramel can usually be distinguished by the naked eye from the dull light brown fragments of coffee, and be removed for examination. The solution of the separate particles in water, and the high colour they impart to it, are indicative of burnt sugar.

When all other forms of adulteration have been disproved, it will be necessary to resort to the estimation of the glucose in the coffee, and to the specific gravity test for the proportion of burnt sugar present. As burnt sugar nearly always contains a considerable proportion of glucose, the detection of any appreciable presents the laticiferous tissue altered by the roasting, and B the woody tissue. The pointed hairs are seen at C, and the sphæraphides at D, whilst E represents portions of fig seeds. The laticiferous and woody tissues, the presence of pointed hairs, and abundance of sphæraphides, form the chief distinctive microscopic characters of figs.

Date Stones.—These have been roasted, ground, and mixed with coffee, and the mixture sold under the name of "Melilotine

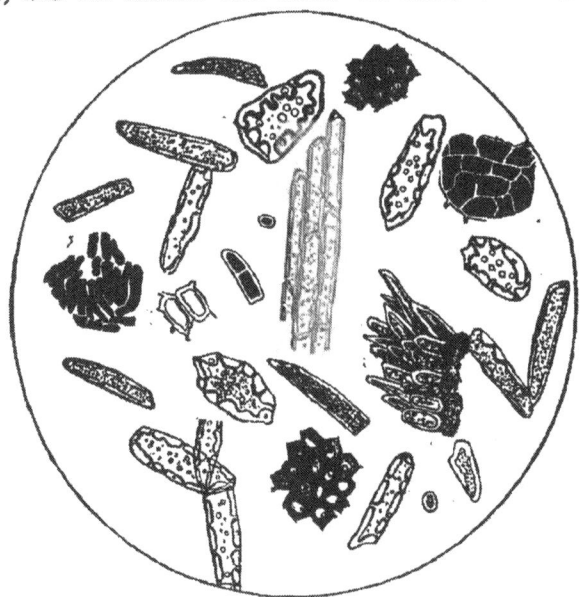

FIG. 22.—DATE STONE.
Magnified 150 diameters.

Coffee." The seed of the date is almost entirely composed of a hard horny substance, the thick-walled cells of which are filled with albuminous and fatty matter.

The microscopic structure of the date stone is very simple. It consists chiefly of sclerogen cells, as shown in Fig. 22. These cells are very characteristic, and can be readily detected and identified in any sample of coffee to which a preparation of date stones has been added.

COFFEE. 69

of small, angular, regularly-formed cells arranged in tolerably regular lines.

The size and form of the cells, the appearance of the starch bags, and the large amount of starch present, render the presence of the acorn easy of detection.

Roasted Figs.—These have lately been imported into this country and sold as a substitute for coffee under the name of "Mochara."

The structure of the fig is rather complex, consisting of cellular, spiral, woody, and laticiferous tissues. In the unroasted

FIG. 21.—ROASTED FIG.
Magnified 150 diameters.

fig these can be readily identified, especially the laticiferous tissue, which is larger and more distinctly branched than that found in any of the roots. When roasted, however, the tissues are more or less altered in character; the spiral and laticiferous tissues being broken up and their internal markings partly obliterated, whilst the woody tissue is rendered more distinct. There are also present in the fig, sphæraphides and unicellular pointed hairs.

The various structures are shown in Fig. 21, where A re-

68 COFFEE.

oval, or oblong cells (seen at A, Fig. 20), with marks or dots on their sides, caused by the minute starch-granules with which the cells are filled. At B these starch-bags are represented more highly magnified, and at C the starch-granules are seen bursting through the cells.

The skin of the acorn is a cellular membrane, containing bundles of spiral vessels, as seen at D, E, and F, the cells being very irregular in form.

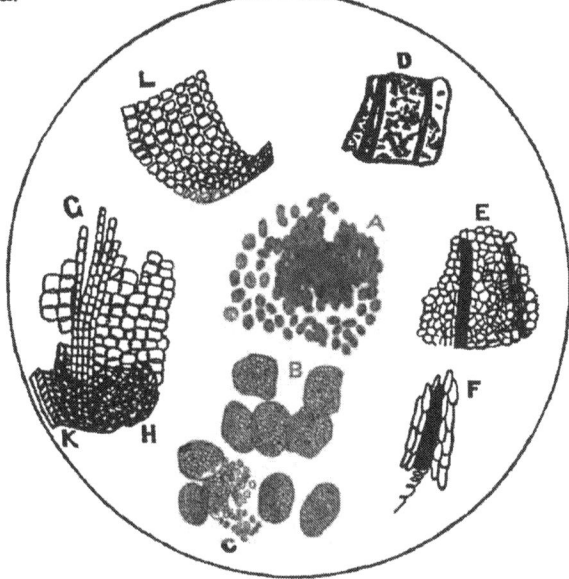

FIG. 20.—ACORN.

Tissues of the Husk and Seed. A, D to K magnified 50 diameters, B and C magnified 150 diameters.

The shell of the acorn, which is but seldom ground up with the seed, consists of two parts, an inner and an outer. The former of these is soft and easily broken up, and the latter hard and horny, so much so, in fact, as to resist all ordinary attempts to disintegrate it. The cells of the inner surface, shown at G, are oblong, and generally run in lines or rows. Those of the outer surface, seen at H, are minute and extremely dense, and are covered by a transparent epidermis, K and L, which is composed

COFFEE. 67

The structure of the seed of the locust bean is delineated in Fig. 19. At A, B, and C, the husk or skin is represented, and at D and E the substance of the seed consisting of thick-walled cells filled with oil-globules and minute starch-granules. These structures are very similar to those usually found in leguminous seeds.

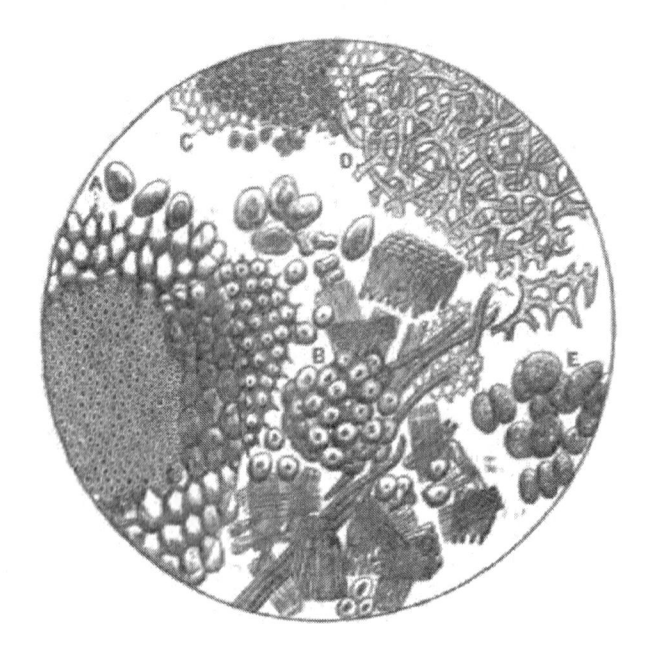

FIG. 19.—LOCUST BEAN.

Tissues of Seed, magnified 150 diameters.

The presence of the wedge-shaped cells, and of starch-granules in the seed, and the very marked structures in the pod, especially the large coloured cells, render the detection and identification of this adulterant comparatively simple.

Acorns.—The seed of the acorn is composed of thin-sided

66 COFFEE.

posed of several very characteristic structures; these are shown in Fig. 18. A represents the epidermis, and B the coloured cells lying underneath, which are easily separated by slight friction. C and E, the third and fifth layers of the pod, consist of woody fibre, amongst which are scattered sclerogen cells. Rows of small round cells are also abundant, lying in a regular manner

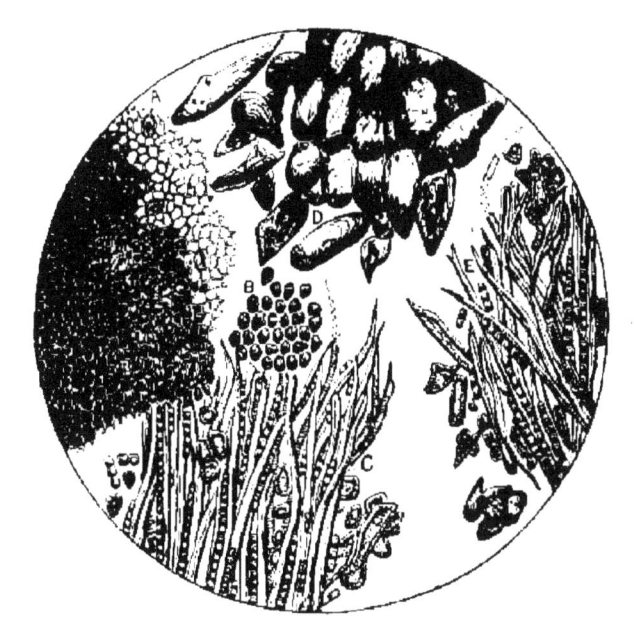

FIG. 18.—LOCUST BEAN.
Tissues of Pod, magnified 150 diameters.

along the sides of many of the woody cells. At D is represented the most characteristic structure of the pod. It consists of large, somewhat oval, cells, having so little cohesion that slight trituration suffices to separate them. These cells have a light brown colour, and are generally striated.

they separate freely from each other, presenting the appearance of tough bags. The starch-granules are also oblong, and have a slit on one side resembling somewhat a grain of barley in miniature. The husk or skin of the bean is a dense thick integument, composed of transparent narrow cells, placed parallel with each other and perpendicularly to the seed. They are represented at A in Fig. 17. The tenacity existing amongst these cells is not easily overcome, but after long boiling with a few drops of caustic potash, the cells may be separated by a little pressure, and they then present the appearance of narrow transparent wedges with light brown centres, as seen at B. In addition to these, there occur other cells having somewhat the shape of dumb-bells; they are represented at C, and are highly characteristic. The inner skin of the husk consists of loose oval cells, as at D, each enclosing what seems to be a second cell, the interior of which is filled with dark granular matter.

Peas.—The structure of the pea is so nearly akin to that of the bean as to render it difficult to distinguish between them by means of the microscope. The skin is a tough thin integument, consisting of minute flat hexagonal cells resembling honeycomb, and resting upon a delicate colourless membrane composed of fine thin-sided cells somewhat larger than the hexagonal cells.

Drawings of the starches of beans and peas will be found in Part II. The wedge-like and dumb-bell-shaped cells, together with the starch-granules, furnish characteristics by which the presence of beans or peas in any sample of coffee thus adulterated can be with certainty identified.

Wheat, Barley, Rye, and Maize.—The microscopic structure of these and other cereals which are sometimes used as adulterants will be found and described in Part II.

Locust or Carob Beans.—The pod of the locust bean is com-

64 COFFEE.

characteristic tissue by which the seed can be identified; but where it happens that the seed is skinned before being roasted, it becomes necessary to seek for evidence of its identity, either in the condition of the cells or in the character of the starch, if any be present. It is not, however, always practicable to distinguish very clearly the particular kind of starch present, owing to the broken condition of the granules produced by torrefaction.

FIG. 17.—BEAN.
Tissues of Bean, magnified 50 diameters.

When a few grains of a sample containing a leguminous or other starchy seed are placed under the microscope, and a little iodine solution added, the particles of roasted starchy matter will appear of a more or less distinct purple colour.

Beans.—The garden bean consists principally of very tough thick-walled cells, seen in mass at E, Fig. 17, filled with large starch-granules. These cells are generally oblong, and when crushed corresponding structures in mangold wurzel, except that the cells are more irregular in size, and have no tendency to separate, but hold together with considerable tenacity. The woody tissue, which is present in abundance, breaks up on boiling into bundles of short flakes having a striated appearance.

The absence of starch and the presence of the woody tissue make the identification of turnip comparatively easy.

Carrots.—The cells of the cellular tissue of this root are small, transparent, and thin-sided, and those of the skin or bark are angular in shape and very regular. The pitted tubes occur amongst the cells, and, excepting that they are somewhat smaller, resemble those of turnip and chicory.

Carrots contain starch, the granules of which are very small and round, and in some cases muller-shaped, with distinct central hilums.

Parsnips.—The tissues of parsnips so closely resemble in form and size those of carrots, as to render it almost impossible by these characters alone to distinguish between them when roasted and ground. The cells of parsnips, however, contain a much greater proportion of minute starch-grains than those of the carrot, and there is a greater irregularity in the size of the granules.

Carrots and parsnips differ from chicory in not containing laticiferous tissue, and also from chicory and the other root adulterants in containing starch.

II.—Seeds.

Seeds are usually best detected by the microscope, and, as neither they nor their seed-coats contain pitted tissue, there is little difficulty in distinguishing them, even when roasted and ground, from roots.

When a seed is roasted and ground with its seed-coat, there is usually to be found in the structure of the latter some percentage of such sugar by the copper test would be confirmatory of the presence of caramel.

MICROSCOPIC EXAMINATION.

In the detection of the adulterants of coffee, the aid of the microscope is always resorted to. It becomes, therefore, a matter of great importance to gain some clear knowledge of the structure of the tissues of roasted coffee, and also of the various substances which are likely to be employed in its sophistication.

In coming to a decision as to the kind of adulterant which may be present in coffee, or any other article in which the microscope is relied upon for detecting the presence of a foreign substance, it is not enough, usually, to trust to any mere drawing or representation of the adulterant indicated. It is always advisable to examine under like conditions a portion of the particular seed or root which it is supposed has been added, and when the percentage is required, to make up mixtures representing different degrees of adulteration, and to judge, from the number of foreign particles on the slide, the extent to which they have been added to the pure coffee or other article.

In some instances where a foreign substance is shown to be present by the microscope, and it is possible to identify it, the knowledge thus obtained becomes of the utmost value in indicating the particular chemical and other tests which should be applied in order to determine the proportion in which the adulterant is present.

I.—Roots.

All the roots that can be employed with advantage for mixing with coffee are very similar in structure. They mainly consist of a mass of cells, among which occur bundles of jointed tubes, the sides of which present the appearance of being marked with bars or pits.

60   COFFEE.

When moistened with water the particles become soft, while those of coffee remain hard and difficult to bruise.

Chicory.—The tissues composing chicory-root are shown in Fig. 14. A represents the pitted tubes, which are present in considerable quantity and are highly characteristic. They resemble

FIG. 14.—CHICORY.

(A) Pitted tubes	...	...	magnified 80 diameters.
(B) Laticiferous tissue	...	...	,, 50 ,,
(C) Cellular tissue	...	...	,, 50 ,,
(D) Skin	...	...	,, 50 ,,

long jointed cylinders, and are distinctly marked with bars or pits. The joints frequently overlap each other, and give to the cylinders an uneven appearance. B represents what are called laticiferous or milk vessels. They are narrow, transparent, branching tubes, filled with latex. In the roasting of the root the latex is dried up, and the interior of the tubes presents the appearance of being filled with granular matter, as shown in the drawing.

COFFEE. 61

c represents the cellular tissue, which consists of round or oval, smooth, and thin-sided cells. Those composing the bark or skin are somewhat angular, as seen at D.

Chicory is free from starch.

The size of the dotted tubes, the peculiar appearance of the

FIG. 15.—MANGOLD WURZEL.

(A) Pitted tubes ... ... magnified 65 diameters.
(B) Cellular tissues ... ... ,, 150 ,,
(C) Skin ... ... ,, 50 ,,

joints, and the presence of the laticiferous vessels, afford a ready means of distinguishing chicory from the other root adulterants. Dandelion root has a structure somewhat similar to that of chicory, but the pitted tubes of the former are narrower and more regular, and the laticiferous vessels more abundant and better defined than the corresponding tissues of the latter.

Mangold Wurzel.—The microscopic appearance of the tissues composing mangold wurzel is shown in Fig. 15. The pitted tubes represented at A are very abundant and large. Many of them have blunt oblique ends, and on trituration they break up into shorter tubes which terminate obtusely. The cells of the tissue shown at B are larger than those of chicory; they are very transparent and thin-sided, possessing little cohesion, the slightest trituration causing them to separate. The cells composing the outer skin are angular.

FIG. 16.—TURNIP.
Tissues of Turnip, magnified 150 diameters.

The absence of laticiferous tissue and the size of the cells serve to distinguish mangold wurzel from chicory. Like the latter, however, mangold wurzel is free from starch.

Turnips.—The turnip when roasted and ground presents the appearance shown in Fig. 16. It is composed of cellular, pitted, and woody tissues. The first two of these are very similar to the

Dates.—Recently these have been prepared and mixed with coffee, or with coffee and chicory, and sold to the public under the name of "Date Coffee."

The pericarp, or portion external to the stone, is composed of cellular and spiral tissues. The cellular tissue partly consists of small well-defined sclerogen cells, which are characteristic, and can be readily recognised. Date Coffee also contains the tissues of the stone shown in Fig. 22.

COCOA.

Botanical Origin.—The cocoa of commerce is prepared from the seeds of the plant *Theobroma Cacao,* which belongs to the natural order *Byttneriaceæ*. The term *Theobroma* implies "food for the gods," and the name was given to the plant by Linnæus, who is said to have had a great liking for the beverage prepared from cocoa. The Mexicans called it "cacaoa quahuitl," naming the preparation from its seeds "chocolatl," and there is no doubt we derive the words cacao and chocolate from these native names. The term cacao has been changed into the familiarly known word cocoa.

The genus *Theobroma* contains a number of species, all of which are natives of the tropical parts of America, and the seeds of several of them are found in the cocoa of commerce. The most important, however, is *Theobroma Cacao,* which yields the best and finest seeds, and which is extensively grown in the West India Islands, Brazil, and Guiana; its cultivation has also been introduced into some parts of Asia and Africa.

The ordinary height of the tree ranges from 12 to 20 feet, but is sometimes much higher. The tree bears leaves, flowers, and fruit all the year round; but although mature and unripe fruit may be seen growing at one and the same time, the chief seasons for gathering the fruit are June and December. The plant, when three years old, begins to bear fruit, but it does not reach its full productiveness till the seventh or eighth year. The leaves, which grow principally at the top of the tree, vary from 7 to 10 inches in length, and from $2\frac{1}{2}$ to $3\frac{1}{2}$ inches in breadth.

Description.—The seeds of the cocoa tree are contained in a pod or fruit, which in shape is intermediate between that of a melon and a cucumber, and measures from 5 to 12 inches in length, and about 3 inches in diameter. The rind of the fruit is smooth, fleshy, and about half an inch in thickness, with a yellow and red tinge on the side next the sun. The seeds are arranged in five rows divided from each other by partitions, and embedded in a pap-like pulp or spongy substance, which separates from the rind when the fruit is ripe, and has a sweet with a slightly acid taste. The number of seeds contained in the fruit is variable, and sometimes amounts (particularly in that grown in Central America), to upwards of forty. The fruit grown in the West Indies and some other parts is much smaller and contains fewer seeds.

The seeds are removed from the recently-gathered fruit, placed in heaps and covered with leaves, or put in boxes or pits in the ground for four or five days, until a certain degree of fermentation has taken place. They are then dried in the sun or before a fire, by which process they become dark in colour, and lose much of their peculiar, harsh and unpleasant taste.

In preparing the cocoa beans for dietetic use, they are roasted in the same way as coffee, in large revolving cylinders over charcoal fires. The main object in the roasting is to develop the greatest possible amount of aroma, and produce a cocoa of full flavour and character. In the process of roasting the loss of weight is estimated at 10 per cent.

After roasting, the beans are passed through a machine which gently cracks the kernels, reduces them to small fragments or nibs, and disengages the husks or shells, which thereby become easy of separation. This separation is effected by passing the crushed beans through a powerful winnowing machine, when the nibs pass out at one end and the broken husks at the other. The nibs are afterwards subjected to a further process of winnowing in small hand-sieves, the object of which is to separate the short steely needle-like germs, and at the same time to remove by the hand all mouldy or discoloured fragments. The sifting in this case is done in small quantities at a time, in order that the whole of the nibs may be exposed to view, and that no objectionable portion may escape detection.

Nibs are the purest form in which cocoa is supplied to the public, but they require long boiling to effect their disintegration and to prepare an extract for use as a beverage. The beans or nibs, when reduced to a powder or paste, can be much more easily cooked, and it is in the ground state in which the great bulk of the cocoa, either mixed or unmixed with other substances, is prepared and sold to the public.

Of manufactured cocoas the two leading types are flake and rock cocoas. Flake cocoa is generally prepared from the entire roasted beans, which are ground in steel mills somewhat similar to those used for grinding coffee, and afterwards reduced to the condition of a coarse paste. Rock cocoa is prepared from the nibs ground to a smooth paste in a heated stone mill, the paste, while in a soft condition, being thoroughly mixed with certain proportions of sugar and some starchy substance such as arrowroot. In some instances sugar only is mixed with the cocoa paste to form rock cocoas.

Most of the other preparations of cocoa, whether sold as soluble cocoa or chocolate, consist of mixtures of cocoa nibs with various substances ground together into a smooth paste. In the manufacture of soluble cocoas, arrowroot, sago, or some other starch, and sugar, either dry or in the form of a syrup, are combined with the cocoa-nib paste. The admixture of starch with the cocoa paste tends to mask the presence of the fat, and to render the cocoa more readily miscible with boiling water. The so-called soluble cocoas, although composed generally of the same ingredients but in different proportions, are sold under various names, such as Homœopathic, Iceland Moss, Rock Cocoa, etc.

The finer chocolates are composed of a mixture of pure cocoa paste and sugar, flavoured with vanilla or some other aromatic substance, but the cheaper kinds contain, in addition to sugar, a certain proportion of starch. Chocolates for eating are prepared with large proportions of sugar and various flavouring substances, and some of these form palatable and nutritious confections.

Cocoa is sometimes deprived of a portion of its fat, and in that state agrees better with many persons than the pure cocoa. The fat is removed by placing the cocoa nibs in bags and subjecting them to pressure in heated presses.

History.—Cocoa appears to have been known to the inhabitants of Central America from time immemorial, but beyond the fact that the Spaniards, at the date of their first visit to the great western continent, found it a common article of food among the natives, nothing definite is known concerning the place which it occupied in their dietary. Though the Spaniards at first disliked the beverage prepared from cocoa, yet in a short time they became exceedingly fond of it, and, its use spreading in their native country, cocoa soon became one of the most valued exports from their American colonies. It is stated that the Mexicans, in addition to preparing their common beverage from cocoa, were in the habit of employing the seeds as a kind of money, and this practice was maintained even at the time of Humboldt's travels in Mexico. Columbus appears to have brought home some samples of the cocoa fruit. For many years the Spaniards imported the cocoa in its prepared state, but ultimately they obtained the necessary knowledge and experience, and commenced its manufacture in Spain from the beans imported in their natural state. This knowledge they carefully guarded, and were thus, for a long time, able to maintain a monopoly of the trade.

It appears that cocoa, like tea, on its first introduction into Europe was considered to possess many valuable medicinal qualities, and that the physicians of the day prescribed it for various diseases. About the beginning of the seventeenth century cocoa found its way into England, and continued to be imported in the prepared state until the beginning of the eighteenth century, when several English manufactories were started. From that time the use of cocoa in this country has continued gradually to increase, as will be seen from the following figures, which exhibit the steady increase of cocoa entered for home consumption since 1820.

1820	267,321 lbs.
1830	425,382 ,,
1840	2,645,470 ,,
1850	3,080,641 ,,
1860	4,583,124 ,,
1870	6,943,102 ,,
1880	10,566,159 ,,

CHEMICAL COMPOSITION.

The cocoa bean contains the following substances: fat, a volatile principle, starch, albumin, astringent matter, gum, cellulose, theobromine, another alkaloid resembling theine, water, mineral matter, cocoa red, and other undefined products of the fermentative action to which the cocoa has been subjected.

We have obtained the following results from the analysis of a sample of raw Trinidad cocoa nibs:

Moisture...	5·23	per cent.
Fat	50·44	,,
Starch	4·20	,,
Albuminous Matter, soluble	6·30	,,
,, ,, insoluble	6·96	,,
Astringent principle	6·71	,,
Gum	2·17	,,
Cellulose	6·40	,,
Alkaloids	0·84	,,
Cocoa Red	2·20	,,
Indefinite Organic Matter, insoluble	5·80	,,
Ash	2·75	,,
	100·00	

Cocoa contains from 6 to 8 per cent. of certain substances which cannot be classed under the better-known compounds mentioned above. They exist partly as products of fermentation, and partly as indefinite vegetable substances, the exact composition of which has not been determined.

We have in the cocoa bean a remarkable combination of nearly all the substances which constitute a perfect food. The non-fatty portion of the cocoa is as rich, if not richer than tea, in total nitrogen, alkaloids, and albuminous matter; and, in addition, it contains a certain quantity of starch, and nearly twice the quantity of important soluble mineral salts found in tea extract. It is not, therefore, strange that cocoa holds so high a place in popular favour.

Exaggerated ideas, we think, are, however, common regarding it, which arise from looking upon it as replacing so much of our more solid articles of food. Cocoa, as generally used, is essentially a stimulating and refreshing beverage like tea and coffee, the addition of any solid, like starch, except as a means of presenting it in a convenient and agreeable form, being unnecessary. A healthy adult who was restricted to the small quantity of solid matter derived from cocoa in a cup of the prepared article, would fare poorly without the addition of a considerable quantity of the usual breadstuffs, meat, etc. If regard be had to the total importation of this article, it is obvious that cocoa forms but a very small proportion of the food of the population, and bears but a very indifferent proportion to the amount of food necessary for each person.

The portion boiled out with water from pure cocoa powder contains by far the more valuable constituents of the bean. It contains nearly all the mineral matter, alkaloids, soluble albuminoids, carbo-hydrates, and astringent substances, all of which are much more likely to be easily assimilated than the part of the bean insoluble in water. This latter is equal to about 20 per cent. of the weight of the bean, and, in round numbers, consists of

7 per cent. insoluble albumen, 7 per cent. cellulose, and 6 per cent. indefinite colouring matter. If pure cocoa powder, therefore, were used in preparing the beverage by merely boiling in water and simply straining, the deficiency arising from the absence of the undissolved part of the cocoa, in a breakfast-cup of the prepared solution, would be more than replaced by less than half an ounce of bread.

Fat.—This is sometimes called cocoa butter. It forms nearly one-half the weight of the cocoa bean, and is obtained as a solid, white, and inodorous fat, which does not readily become rancid. It is soluble in boiling alcohol, from which it crystallizes on cooling; it melts at about the temperature of butter-fat, 86° F. (30° C.), and yields, on saponification, oleic and certain other fatty acids, about which little is known.

Volatile Principle.—This is said by some to be an essential oil. It is doubtful whether this is the case, as no trace of the characteristic odour of cocoa is obtained from the fatty residue removed from the cocoa by exhaustion with benzol. The volatile aromatic principle is dissolved out by cold water, and can from this extract be distilled over in presence of water, but no evidence of an oil is obtainable. It is no doubt produced during the fermentative process to which the cocoa is subjected, and further modified by the subsequent roasting of the beans.

Starch.—This is a characteristic though not very abundant constituent of cocoa. The amount of starch present appears to have been over-estimated by several chemists who have investigated cocoa. The cocoa nib contains from 4 to 5 per cent. of starch, or about 8 per cent. calculated on the non-fatty portion of the bean.

Albumin.—This is present to a great extent in the insoluble form, and is associated with the cellulose in the portion of the nib left after continued boiling in water. As in the case of tea and coffee, an indefinite albuminous compound is dissolved out from the powdered cocoa by boiling alcohol. This albuminoid principle may not improbably have a more active and nourishing property than the substance usually recognised as ordinary albumin.

Astringent Principle.—This substance appears to have been but little studied. It is of the nature of a tannin, though differing much from the tannin of tea and coffee. It rapidly changes during the process of analysis into cocoa red, which renders its quantitative estimation a matter of difficulty. It is precipitated by subacetate of lead, and can be obtained after removal of the lead by sulphuretted hydrogen. It gives a green precipitate with ferric chloride not unlike caffeic acid under similar treatment. When heated with a little chlorate of potash it undergoes a change, and then gives, with ferric chloride, a purple-black colouration, and, like tannic acid, is precipitated by gelatin. The astringent matter also gives an appreciable amount of glucose on being boiled for some time with a mineral acid.

Gum.—A very small quantity of a gum-like substance is obtained from the cold-water infusion of the cocoa nibs. It somewhat resembles dextrin, but from the smallness of the amount there is but little importance attaching to it.

Cellulose.—Cocoa contains from 6 to 7 per cent. of a substance corresponding to cellulose. Few attempts appear to have been made by chemists to estimate this body as distinct from starch, both substances being generally converted into sugar, and the total amount obtained reckoned as starch. Hence the amount of starch shown in some analyses of cocoa is nearly twice the actual quantity present.

Theobromine.—$C_7 H_8 N_4 O_2$. This substance is the principal alkaloid of cocoa. It is found in nearly equal quantities in the husk and "nib" or kernel of the cocoa bean. It was discovered by Woskresensky, in 1841. Payen found 2·0 per cent.; Mitscherlich, 1·5 per cent.; Playfair and Lankester, 2 per cent. Other chemists show much smaller amounts: Tuchen, 0·38 to 0·66; Hassall, 0·47 to 0·78; and Muter, 0·9 per cent.

We have obtained the following percentages of theobromine,

and of a theine-like alkaloid, from the subjoined varieties of cocoa nibs:

Cocoa.	Theobromine per cent.	Theine-like Alkaloid per cent.
Guayaquil	0·54	trace
Grenada	0·91	,,
Surinam	0·78	0·02
Trinidad	0·59	0·25
,, Husk	1·02	0·33

Theobromine, when quite pure, crystallizes in well-defined rhombic prisms. It is almost insoluble in benzol and petroleum spirit, very slightly soluble in cold water, alcohol, and chloroform, but easily soluble in boiling water and alcohol. It contains 31·1 per cent. of nitrogen, being richer in this substance than theine, which it otherwise closely resembles. Its subliming temperature has been fixed at about 554° F. (290° C.), but it has been found that some of the theobromine is volatilised below that point. It forms, with acids, well-defined crystalline compounds, among which may be mentioned the nitrate and hydrochlorate. Like theine, it forms murexide by the action of nitric acid and ammonia.

The alkaloid which has been found associated with theobromine crystallizes in fine silky needles resembling theine, with which it also corresponds in its solubility in benzol. It is present in both kernel and husk—in Trinidad cocoa to the extent of 0·25 per cent. of the kernel, and 0·33 per cent of the husk; but Surinam nibs yielded only 0·02 per cent. of this substance. It contains 25·48 per cent. of nitrogen.

Cocoa Red.—This substance gives the characteristic colour to cocoa. It does not appear to be present in the freshly-gathered bean, but arises from the oxidation of the astringent principle or natural tannin of the cocoa berry. This may account for the variable proportions of this substance found in different cocoas. It has the character of a resin, and is partly soluble in hot water, but more soluble in alcohol. The different degrees of solubility may probably coincide with the extent of oxidation it has undergone. A portion of this substance remains insoluble in the cocoa, even after treatment with boiling alcohol and water.

Ash.—In four descriptions of cocoa-nibs we have found the percentage of ash to vary from 4·47 to 5·71, but the amount of inorganic matter in the husk is much higher. In the husk of the finest Trinidad beans we have found 10·19 per cent.

A complete analysis of the ash of cocoa will be found on page 86.

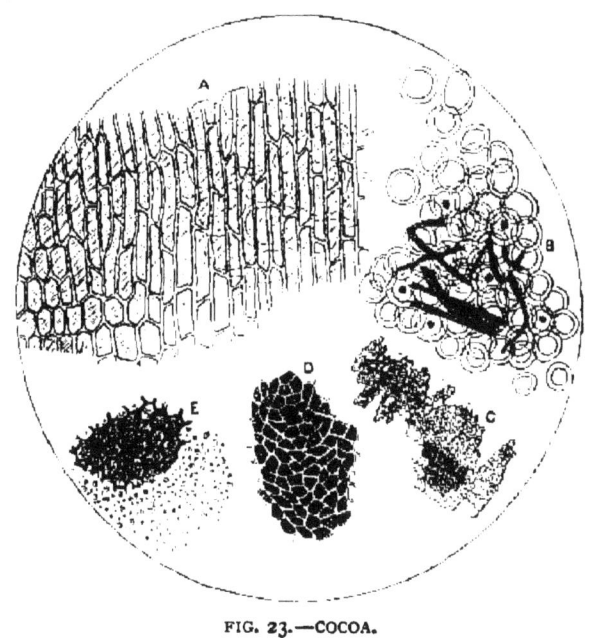

FIG. 23.—COCOA.

MICROSCOPIC STRUCTURE.

The cocoa bean has been already described as consisting of a husk or shell, and a kernel, the latter, when broken into fragments, forming the cocoa nibs of commerce. In Fig. 23, at A, B, and C, is given the microscopic structure of the tissues of the husk, and at D and E that of the tissues of the kernel.

The outermost tissue of the husk represented at A consists of two layers of elongated quadrangular cells, one of which layers lying upon the other in a slanting direction imparts to the tissue a striated appearance. Towards the ends of the bean the cells of this tissue become smaller and somewhat hexagonal. The second tissue, seen at B, constitutes the main substance of the husk, and consists chiefly of large circular or slightly angular cells containing mucilage and a considerable number of sphæraphides. Several rows of spiral vessels run through this tissue from end to end. The third tissue, represented at C, is very characteristic, and consists of small hexangular cells, some of which are filled with dark granular matter.

The kernel consists of a number of lobes irregular in size and shape, each of which is enveloped in a membrane composed of angular cells filled with granular matter, as seen at D. The substance of these lobes consists of a mass of hexangular cells, represented at E, containing small round starch granules, having a central hilum.

ANALYSIS.

Fat.—This is best estimated by the use of petroleum spirit, or benzol. Fifty grains of the dry cocoa are bruised in a mortar, with the solvent employed, and the dissolved fat carefully separated by the aid of a good filter-paper. This operation is repeated several times, or until all traces of fat are removed. On the evaporation of the benzol, the fat is obtained as a perfectly white residue. We have obtained by this means from 47·0 to 53·0 per cent. of fat from the cocoa nibs.

Starch.—In the estimation of the starch and other constituents of the cocoa, the non-fatty portion is used which is insoluble in the benzol and left as an almost impalpable powder. 50 grains of this powder are exhausted with alcohol; the insoluble residue is then dried and boiled in water for four hours, or until all the starch is rendered soluble. The portion insoluble in boiling water should not give any colouration with iodine. The dissolved starch is boiled for 6 hours, with 20 cubic centimetres of normal sulphuric acid. Basic acetate of lead is then added, the filtrate decomposed with sulphuretted hydrogen, and the sulphide of lead removed by filtration. The filtrate or solution is next made up to a given bulk, and the sugar formed by the action of the sulphuric acid on the starch estimated by Fehling's solution. From the proportion of sugar obtained the amount of starch is calculated. As cocoa contains matter which very readily reduces copper solutions, the precipitation with basic acetate of lead must always be resorted to. Any estimation of sugar performed without this precaution will give fallacious results.

Albumin.—The nitrogen in the portion insoluble in boiling water is estimated by combustion with copper oxide, and the percentage thus obtained multiplied by 6·3, gives the amount of albumin. The proportion of nitrogenous matter thus obtained is only about one-half the entire quantity present in the cocoa bean, exclusive of theobromine; the remainder is dissolved out by alcohol and boiling water. The amount of nitrogen in the soluble portion is also estimated by copper oxide, but its conversion into an equivalent of albumin by the usual formula may, in the uncertainty attaching to the composition of some of these nitrogenous bodies, give rise to an erroneous result.

Astringent Matter.—This is estimated by digesting the non-fatty cocoa first with alcohol and then with boiling water. Keeping the two filtrates distinct, the latter is boiled down to a small bulk and alcohol added to it. Any precipitate that falls is filtered off, and the filtrate added to the primary alcoholic solution. To the mixed alcoholic solutions is next added acetate of lead, and the precipitate that then falls is separated, washed, diffused in water, and the lead precipitated by passing a current of sulphuretted hydrogen through the liquid and filtering off the lead sulphide. The clear and colourless filtrate is then evaporated to dryness, and the residue weighed, the ash being subsequently determined and deducted from the total dry residue. As the solution evaporates it assumes a bright red hue, and part of the colouring matter of the residue becomes insoluble in water.

Gum.—Five grams of the cocoa are exhausted first with benzol and then with alcohol. The residue is next treated with cold water for two hours, and filtered; the aqueous filtrate is evaporated to a small bulk, filtered, and to the filtrate alcohol of 90 per cent. is added to precipitate the gum. The precipitate is thrown on a filter, washed with alcohol, and re-dissolved in a small quantity of water. The gum is again precipitated with alcohol, dried, and weighed on a tared filter. The amount of mineral matter in the gum is determined by ignition, and deducted from the total weight.

Cellulose.—After the starch has been rendered soluble, and the solution filtered off, the residue is boiled with 20 cubic centimetres of normal sulphuric acid for six hours, and the resulting sugar estimated with the precautions observed in the case of starch. In addition to the direct estimation before described, the starch may be determined by difference—that is, by deducting the amount equivalent to the cellulose from the total sugar formed after conversion of the starch and cellulose in the part insoluble in alcohol.

Theobromine.—One of the earliest methods adopted for the separation of this alkaloid very much resembles that employed for the estimation of caffeine. An alcoholic extract of the bean was made, the alcohol evaporated, basic acetate of lead added, and the precipitate separated. Sulphuretted hydrogen was then passed through the filtrate, the precipitated sulphide of lead separated, and the solution evaporated nearly to dryness. Finally, the mass was extracted with alcohol, the filtrate partly evaporated, and then left to crystallize. On repeating the crystallization the theobromine was obtained as an almost white powder.

Our analysis of cocoa has confirmed the presence of a second distinct crystalline body, which exists in larger quantities in some descriptions of cocoa than in others, and in still larger proportion in the husk than in the corresponding kernel. As these alkaloids possess different degrees of solubility in the re-agents usually employed for the estimation of theobromine, it is necessary to employ certain precautions, so that the total amount of alkaloid may be obtained. The following process has been applied by us for the separation of the two substances, which were obtained in a state of considerable purity.

One hundred grains of cocoa were repeatedly digested with hot benzol, and filtered. The benzol was distilled from the filtrates, the resulting fatty residue boiled with water, and then allowed to become cold, and the aqueous solution filtered off. This process was twice repeated to ensure complete removal of the alkaloid dissolved out by benzol. The water extract was then evaporated to dryness, and the residue purified by solution successively in water and in benzol. A substance was then obtained—especially from Trinidad cocoa—which crystallized in fine silky needles, after the manner of theine.

The cocoa which had been treated with benzol was mixed with 100 grains each of calcined magnesia and sharp sand, and made into a paste with water, and well mixed in a mortar. The mixture was dried in a water-bath, and afterwards repeatedly digested with separate quantities of alcohol of 90 per cent. strength for fifteen minutes in a flask with a cohobating apparatus attached. The filtrates were collected and the alcohol distilled off. The residue which contained the theobromine was transferred to a capsule, dried, and weighed. The dry residue was then treated with hot benzol to remove any traces of fat and of the theine-like crystals, which were subsequently isolated and weighed. The residue insoluble in benzol was then twice treated with a little water, which had been cooled below 40° F. (4·4° C.), and on decanting the water the theobromine was left as a powder perfectly white and pure, with the exception of a trace of inorganic matter, which was estimated by ignition. As even the cold water in some instances dissolved out a little of the theobromine, it was found desirable to evaporate the water extract to dryness, and again treat with cold water, when a further small quantity of crystals was obtained. If the crude residue obtained by alcohol from the magnesia mixture were submitted to a nitrogen combustion with copper oxide, and the nitrogen found calculated into theobromine, a higher and probably more accurate percentage of alkaloids would be obtained.

Ash.—The following table exhibits the percentage composition of the ash of cocoa nibs and husk:

ANALYSIS OF THE ASH OF COCOA NIBS AND HUSK.

Constituents.		1. Guayaquil Nibs.	2. Surinam Nibs.	3. Grenada Nibs.	4. Finest Trinidad Nibs.	5. Finest Trinidad Husk.
Sand		—	—	—	—	5·12
Silica		·15	—	—	—	2·87
Chloride of Sodium	NaCl	·46	·53	·57	·65	·44
Soda	Na_2O	·46	·63	·57	·83	·94
Potash	K_2O	23·35	28·00	27·64	29·30	37·89
Magnesia	MgO	19·18	20·66	19·81	18·23	13·04
Lime	CaO	3·24	4·38	4·53	6·51	7·30
Alumina	Al_2O_3	·10	·04	·08	·08	·55
Protoxide of Iron	FeO	·21	·38	·15	·10	·63
Carbonic Anhydride	CO_2	·69	3·31	2·92	4·19	10·80
Sulphuric ,,	SO_3	2·77	4·29	4·53	3·91	3·25
Phosphoric ,,	P_2O_5	49·39	37·78	39·20	36·20	17·17
Total		100·00	100·00	100·00	100·00	100·00

It will be seen from the above analyses of the ash of the cocoa nib that the total amount of ash and of the several constituents of which it is composed, very closely correspond in the case of Nos. 2, 3, and 4. No. 1—the Guayaquil—has a pre-eminent position with regard to its total ash, and perhaps also in an equal degree in the amount of its phosphoric acid (P_2O_5), which forms practically one half of the entire ash and is equal to 1·79 per cent. on the whole cocoa kernel. The ash is distinguished by the small amount of chlorine, soda, and carbonates present, and it may also be remarked that the magnesia is from three to five times greater in quantity than the lime, a circumstance somewhat unusual in the ash of vegetable products.

The ash obtained from the husk is about three times the quantity found in the kernel of the cocoa from which the husk was derived. Though the phosphoric acid bears a lower proportion to the other constituents of the ash of the husk, yet in actual amount it is half as much again as the phosphoric acid in the same weight of the kernel. The potash alone in the husk is more than the average weight of ash obtained from the four varieties of the kernel.

TABLE SHOWING THE PERCENTAGE OF MOISTURE, AND OF TOTAL ASH IN VARIOUS COCOAS, WITH THE RELATIVE SOLUBILITY OF THE LATTER.

Description.	Per Cent. Moisture.	Soluble in Water.	Insoluble in Water, but soluble in Dilute Hydrochloric Acid.	Insoluble in Dilute Hydrochloric Acid. (1 to 8).	Total.
Guayaquil Nibs	5·06	2·04	1·59	none	3·63
Surinam ,,	4·55	1·26	1·64	,,	2·90
Grenada ,,	5·71	1·37	1·45	,,	2·82
Finest Trinidad Nibs	4·47	1·28	1·47	,,	2·75
,, ,, Husk	10·19	4·74	3·38	0·51	8·63

COCOA.

Table showing the Percentage of Ash Soluble and Insoluble in Water and Dilute Hydrochloric Acid respectively.

Description.	Soluble in Water.	Insoluble in Water, but soluble in Dilute Hydrochloric Acid.	Insoluble in Dilute Hydrochloric Acid. (1 to 8).
Guayaquil Nibs	56·20	43·80	none
Surinam ,,	43·45	56·55	,,
Grenada ,,	48·58	51·42	,,
Finest Trinidad Nibs	46·55	53·45	,,
,, ,, Husk	54·92	39·17	5·91

A noteworthy feature in connection with the mineral matter of cocoa is the large proportion which is soluble even in cold water. Two cocoas were exhausted with cold water and the mineral matter in the filtrate estimated by ignition, with the following results :

    Guayaquil ... ... 2·96 per cent.
    Trinidad ... ... 2·44 ,,

It will thus be seen that, in using only the boiled extract of cocoa for dietetic purposes, nearly all the available mineral matter is obtained in solution.

ADULTERATION OF COCOA.

The practice has existed for many years of adding starch and sugar to cocoa in the manufacture of prepared or soluble cocoas, and so long as these preparations are not sold under the denomination of pure or unmixed cocoa, the admixture cannot strictly be called a form of adulteration.

As the proportion of fat in the natural cocoa has been found to disagree with many persons, a considerable popular demand has arisen for cocoas less rich in this constituent, and there appears to be little or no objection to the removal of, say, a moiety of the fat, which would still leave, in the case of most cocoas, at least 25 per cent.

The following tables show the general composition of commercial cocoas:

No. 1.
ANALYSES OF COMMERCIAL COCOAS.

Per Cent. of	Prepared Cocoa.	Iceland Moss Cocoa.	Rock Cocoa.	Flake Cocoa.	Cocoatina.	Chocolatine.	Finest Trinidad Nibs.	Chocolate de Santé.	Cocoa Extract.
Moisture	4·95	5·47	2·58	5·49	3·52	4·40	2·60	1·44	5·76
Fat	24·94	16·86	22·76	28·24	23·98	29·60	51·77	22·08	29·50
Starch (added)	19·19	24·70	17·56	none	none	none	none	2·00	none
Sugar (cane)	23·03	29·23	32·20	,,	,,	,,	,,	61·21	,,
Non-fatty Cocoa	27·89	23·74	24·90	66·27	72·50	66·00	45·63	13·27	64·74
	100·00	100·00	100·00	100·00	100·00	100·00	100·00	100·00	100·00
Percentage of Nitrogen	2·24	1·38	—*	3·06	4·07	4·36	2·95	—*	—*

* Nitrogen not determined.

No. 2.
ANALYSES OF COMMERCIAL COCOAS.

Name.	Per Cent. Of Ash.	Per Cent. Of Cocoa soluble in Cold Water.	Amount of Ash in portion Soluble in Cold Water.
Prepared Cocoa	1·52	31·66	1·17
Iceland Moss Cocoa	1·83	40·80	1·06
Rock Cocoa	1·56	36·70	0·90
Flake ,,	5·39	18·10	4·00
Cocoatina	6·81	18·00	3·95
Chocolatine	6·14	18·50	4·50
Finest Trinidad "Nibs"	2·86	10·58	2·44
Chocolate de Santé	1·76	65·60	1·26
Cocoa Extract	5·64	16·72	4·36

From the above analyses it will be seen that several of the commercial cocoas consist of the powdered nibs only deprived of from 40 to 50 per cent. of their fat.

"Flake cocoa" contains the husk in a ground state. As the husk contains more ash and extractive than the kernel, the presence of the former raises the proportion of the ash; and in cases where the quantity of ash is taken as a criterion of the value of a prepared cocoa, the husk gives an apparent increase in value.

The four mixed cocoas in Table 1 fairly represent the composition of the commercial articles of this class.

Some of the results of an analysis of a sample of Trinidad "nibs" are entered in Table 1 as a means of comparison, Trinidad cocoa being selected as, in chemical composition, it stands midway between the poorer and richer varieties of cocoa.

It is said that the adulteration of cocoa has been practised by the addition of chicory, oxide of iron, ferruginous earth, chalk, sulphate of lime, etc. In recent years, however, there is little reason to believe that these substances have been added to cocoa.

In the examination of commercial cocoas the estimation of the fat and of the added starch and sugar will usually give an accurate idea as to the amount of cocoa contained in a sample. An indirect method of arriving at this result, and employed by some chemists, is to make a cold-water extract of the cocoa, estimate the dry extract and percentage of ash contained in it, and, by the adoption of certain standards for pure cocoa, by a simple calculation determine the proportion of genuine cocoa present. This method is beset with some difficulties, and these partly arise from the want of uniformity in the results obtained from different kinds of genuine cocoa, and from the presence of the husk, which gives more extractive and mineral matter to water than the nib, and therefore unduly increases these.

About three grams of the cocoa, and 200 cubic centimetres of water are taken. The cocoa is well rubbed up with a little of the water in a porcelain mortar, and thoroughly rinsed out with the remainder into a flask. The cocoa is left in contact with the water for a definite number of hours, and then a sufficient quantity filtered to allow 40 or 50 cubic centimetres being taken for evaporation to dryness, and from the result the total amount soluble in cold water is easily calculated. The dry residue is ignited to obtain the amount of ash. In the analysis of a Trinidad cocoa, the amount of ash in the cold water solution was 2·44 per cent., and that obtained from a prepared cocoa was 0·90 per cent. The calculation of the proportion of cocoa in the latter sample from the above data is therefore as 2·44 : 0·90 :: 100 : 37. This result is below the true quantity, judging from the direct estimation of the sugar and starch in the sample. The determination of the ash affords a useful but rather rough test.

Sugar.—To directly estimate the sugar in commercial cocoa, two grams are taken and exhausted with alcohol of 70 per cent. strength; the spirit is evaporated, acetate of lead added, and the precipitate separated; the filtrate is freed from lead with sulphuretted hydrogen, and the solution boiled with a few drops of sulphuric acid to invert the cane-sugar, and the amount present is then determined by a standard solution of copper salt.

Starch.—It has already been stated that cocoa contains from 4 to 5 per cent. of starch; but there is no difficulty in distinguishing and identifying the starches commonly employed for mixing with cocoa from the starch natural thereto. The granules of the starch of cocoa differ in size and shape, as well as in other characters, from those of the starches which are used in the preparation of some commercial cocoas. The starches employed for this purpose are limited in number, and usually consist of arrowroot, sago, tous-les-mois, and sometimes potato.

92   COCOA.

Arrowroot.—The microscopic appearance of this starch is represented in Fig. 24. The granules are more or less oblong

FIG. 24.—ARROWROOT.

in shape, and generally marked with a series of delicate concentric rings.

COCOA. 93

Sago.—The appearance of the granules of this starch under the microscope is given in Fig. 25. The granules are elongated,

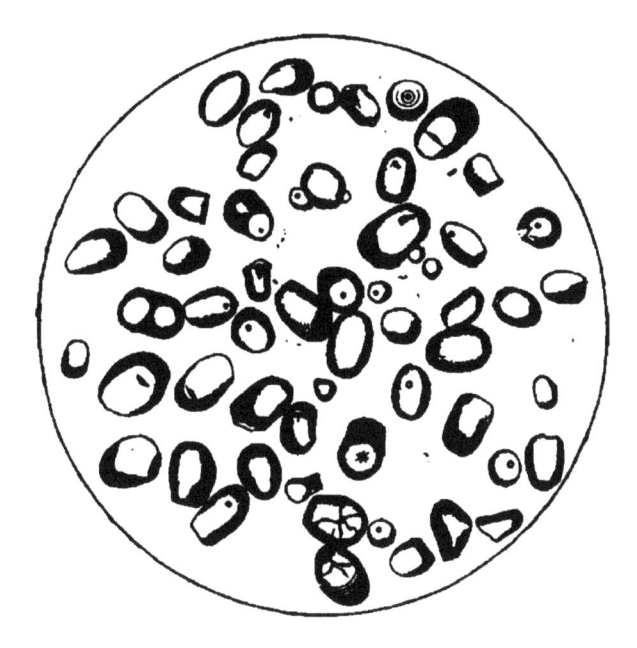

FIG. 25.—SAGO.

round at one end, truncated at the other, and altogether so distinctive in size and shape that they are readily identified.

Tous-les-Mois.—The granules of this starch are given in Fig. 26. They are ovate in form, obtuse at one extremity and somewhat pointed at the other, and are marked by a series of regular and distinct concentric rings. The hilum, which is circular, is situated near the pointed extremity.

Potato.—The granules of this starch are represented in Fig. 27. They vary greatly in size and form; some being triangular or oyster shape, some ovate, and others round, especially the smaller ones. The granules are marked by well-defined concentric rings, and bear a hilum which is sometimes circular and sometimes stellate.

A general idea of the quantity of added starch may be arrived at by the microscope, by comparing the appearance of the mixed cocoa with equal portions of cocoa containing known percentages of the kind of starch identified in the sample.

The amount of added starch in a cocoa may also be determined by chemical analysis. In making this analysis the residue left after treatment with alcohol for the estimation of sugar, is boiled with water for ten minutes with the addition of five drops of sulphuric acid. By this means the starch is rendered soluble and can be readily filtered from the albuminous matter, whilst the cellulose remains practically unaffected. The soluble starch is removed by filtration, and, after the addition of ten drops of sulphuric acid, boiled for six hours to convert the starch into glucose. The colouring matter is removed as usual with subacetate of lead, and the amount of sugar estimated by the copper solution, from which the percentage of starch is calculated. As cocoa contains from 4 to 5 per cent. of starch, an equivalent deduction on this account must be made from the total amount of starch obtained.

Chicory.—The presence of chicory is indicated by the high colour of a cold-water extract, and proved by the subsequent detection of the characteristic tissue by the aid of the microscope. It might be estimated by one of the methods described under Coffee.

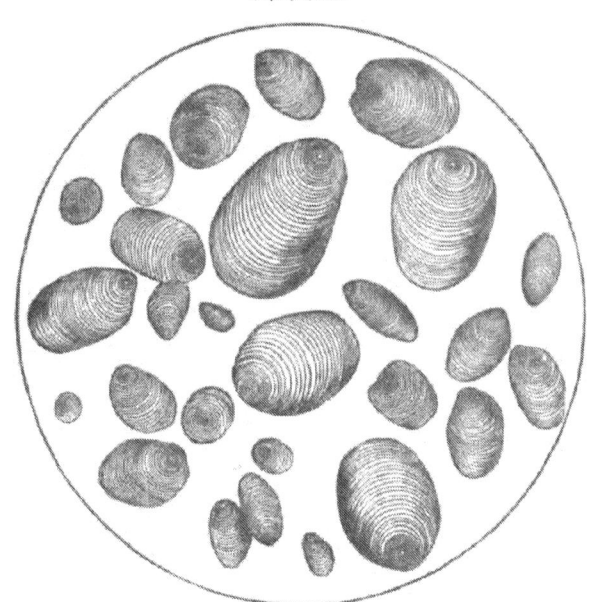

FIG. 26.—TOUS-LES-MOIS.

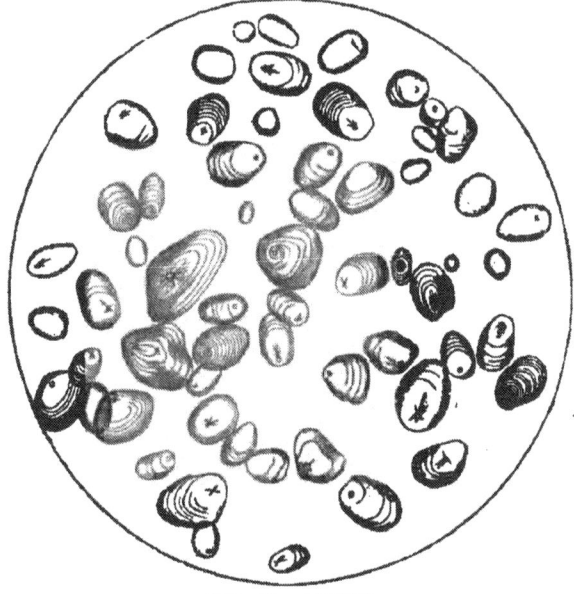

FIG. 27.—POTATO.

Iron Oxide.—This compound is indicated by the red colour of the ash residue after ignition of the cocoa; but its presence, as well as that of the sulphate and carbonate of lime, is of such rare occurrence that comparatively little interest attaches to this part of the analysis. If either of the substances were indicated, their estimation could be effected by the usual process for the estimation of lime, sulphates, etc.

SUGAR.

SUGARS are amongst the most widely-diffused substances in the vegetable kingdom; but those which appear in commerce are capable of being divided into two classes—viz. that corresponding to sugar from the sugar-cane, and the class corresponding to glucose.

Origin.—Sucrose, or cane-sugar, has its source—

1st. In certain grasses, to which the generic name *Saccharum* is applied, and which are cultivated chiefly in India, the East and West India Islands, Mauritius, South America, and China.

2nd. In *Sorghum saccharatus,* a native of India, cultivated in the United States under the name of Chinese sugar-cane, and known also as "Sorghum."

3rd. In the common beet (*Beta vulgaris*).

4th. In the sap of the sugar-maple (*Acer saccharinum*).

5th. In the sap of certain palms, such as the cocoa-nut (*Cocos nucifera*), the wild date-palm (*Phœnix sylvestris*), and the *Borassus flabelliformis,* from which is derived a low-class sugar, known in commerce by the name of "Jaggery."

6th. In the green stalks of maize or Indian corn (*Zea mays*).

The sugar derived from the above sources is that which is almost exclusively used for domestic purposes. It is crystalline, soluble in one-third of its weight of cold water, and, when pure, has a specific gravity of 1·593 at 39° F. (4° C.)

The principal sugars corresponding to glucose are—*Dextro glucose*, produced by the hydration of starch under the influence of dilute acids, and existing ready-formed, sometimes with cane-sugar and sometimes with lævulose in fruits; *Maltose*, produced by the action of malt-extract on starch; *Lævulose*, formed from cane-sugar, along with dextro-glucose, by the action of dilute mineral acids—the two kinds existing in equal proportions in what is known as Invert Sugar; and *Mannitose*, produced by the oxidation of mannite.

Description.—*Cane-sugar.*—There are a great many varieties of sugar-cane or *Saccharum officinarum*, the most important being the common cane, having a yellow stem, and called by the West Indians "Creole," or "Native Cane;" the purple cane, having a purple stem and richer juice, and called in India "karambou kari;" and the gigantic cane, having a large, light-coloured stem, called in India, "karambou valli." They are all the products of tropical or sub-tropical climates, flourishing best in moist and nutritious soils. They vary in height from 6 feet to 18 feet, the average being 10 or 12. The diameter of the stem is from 1½ to 2 inches, and the joints on a single stem number from 40 to 80. They require from twelve to sixteen months to arrive at maturity. After being cut down, they strike again from the stole, the fresh stems being called "rattoons." The length of the joint diminishes every year, while the richness of the juice increases. At the end of five or six years, however, it is necessary to renew the plantation. As soon as the canes are ripe, they are cut down during dry weather, stripped of their leaves and watery top joints, which serve for shoots, then carried as quickly as possible to the mill-house, and crushed between rollers. The juice—which, as it flows from the mill, varies in specific gravity from 1·067 to 1·106—is then heated slightly, and treated either with milk of lime or sulphite of ?, by which the small quantity of acid present in the solution is neutralised, the tendency to fermentation corrected, and a large quantity of impurities made to rise to the top in the form of scum. This being removed, the partly-clarified juice is evaporated down and crystallized, the product being imported into this country chiefly as raw sugar.

Cane-syrups show a greater tendency than beet-syrups to become acid and pass into the invert form of sugar in the process of boiling down, and the loss of crystallized sugar from this cause is very large. It has been estimated that sugar-cane contains about 18 per cent. of sugar, of which about one-third is left in the canes after crushing, and is therefore lost; that an acre of land will grow 30 tons of canes; but that, owing to the loss in crushing and boiling down, the acre does not produce more in actual practice than 2½ tons of raw sugar. The molasses produced is to a great extent used in distilleries for the manufacture of rum. Large quantities of raw sugar imported into this country go into the hands of the refiners, whose business it is to cleanse or purify it. To effect this it is redissolved, generally in about half its weight of water, and the solution sometimes heated with bullocks' blood, which, on coagulation, has the effect of removing solid impurities, the coagulated mass being separated by filtration. Having been again heated, the syrup is run through a long column of animal charcoal, whence, having been partially decolourised and deprived of gummy matter, it is passed to a vacuum-pan, and concentrated until it becomes what is called "supersaturated," after which it is allowed to crystallize. If it is required to make loaf-sugar, the air is admitted when the liquor has got beyond the crystallizing point, and the temperature is raised, by which some of the smaller crystals pass again into solution. This heated mixture of liquid and crystals is thrown as rapidly as possible into moulds, in which it is allowed to stand for a couple of days to drain and harden. At the end of this time a saturated solution of pure white sugar is passed through the moulds in order to wash the crystals and remove any impure syrup from the pores. The sugar is next removed from the moulds and placed in a stove heated by steam, for four or five days, or until thoroughly dry, when it is fit for the market.

Beetroot-sugar.—The process for the manufacture of sugar from beetroot is as follows: The roots are first washed, and then freed from the crowns, as well as from any decayed portion. They are then either crushed or subjected to what is called the "diffusion process." If the first plan be adopted, the roots are placed in a machine containing a large drum, the circumference of which is set with a number of saw-blades or teeth. This drum rotates, and as the roots are passed through they are rasped into a pulp, which falls into a cistern or hopper, where the juice is separated by pressure. In the diffusion process, the cleaned roots are sliced and placed in warm water for a time. The mixture of juice and water thus formed is then run into a second vessel containing more sliced roots, and so on to a third and a fourth, by which time it generally becomes sufficiently concentrated. The juice thus obtained is treated in the same way as the product of the crushing operation—that is, it is first heated with milk of lime to a temperature of 140° to 150° F. (60° C. to 66 C.) in order to clarify it, and, the scum having been removed, carbonic acid is forced into it to precipitate any lime which has been dissolved in the syrup, and in doing which it also carries down with it a quantity of colouring matter. The next operation is to pass the juice through animal charcoal, after which it is put into a vacuum-pan heated by steam, and evaporated to a syrup, having a specific gravity of from 1·200° to 1·300°, and containing from 45 to 60 per cent. of sugar. The methods adopted for obtaining pure sugar from this syrup do not differ essentially from those employed in refining the product of the sugar-cane. The refiner of beet-sugar, however, finds his syrups to contain a much larger quantity of potash salts, which it is his object to leave behind as completely as possible in the mother liquor, from which his last crystallization of sugar has been obtained.

Maple-sugar.—The sugar-maple, *Acer saccharinum*—a tree abundant in the forests of North America—is a valuable source of sugar to the inhabitants in the interior of the Northern States. The tree is tapped in the spring of the year, and the sap which runs from it collected in vessels, and boiled down to the crystallizing point and strained, after which it is poured into moulds to solidify, and is then ready for use.

Jaggery.—The sugar called "Jaggery," the product of various palm-trees, supplies the wants of a large number of the inhabitants of India. It is imported into this country, but is perhaps the lowest quality in the market, and is unfit for domestic purposes in its raw state, and, from the large proportion of invert-sugar contained in it, commands a very low price from the refiner. It however exercises an important influence in controlling the value of raw sugar in times of scarcity.

Molasses, etc.—The mother liquor left in the crystallization of raw sugar is called Molasses; that in the crystallization of refined sugar, Treacle, or Golden Syrup. These vary in composition, but contain some cane-sugar along with invert-sugar. The invention of the vacuum-pan—which is simply a means of boiling at a temperature lower than the boiling-point of the liquid at the ordinary pressure of the atmosphere—has largely reduced the amount of treacle and molasses produced in sugar manufacture. The treacle or molasses from beet is not considered fit for domestic use, and is therefore usually disposed of to the distiller to be converted into spirit.

Glucose.—The production of glucose from grain, or other substances containing starch, is now carried on extensively in this country, and the sugar made is almost exclusively used as a substitute for malt in the brewing of beer. There are at least two processes followed. In the first, the grain, being ground and separated from the husk, is thrown into a mixing-vat with about four times its weight of water, and from 2 to 4 per cent. of its weight of sulphuric acid according to the nature of the substances used. This mixture is then passed into what is called a converter, where it is boiled for from twenty minutes to half an hour under a pressure of about 70 lb. of steam, corresponding to a temperature of about 306° F. (152° C.), by which the starch is changed into glucose or dextrose. Chalk is added to remove the sulphuric acid, and, after filtering, the syrup is partially evaporated in a vacuum-pan, and run through a column of charcoal to decolourise it. The next process is to boil it down to the necessary consistence in a vacuum-pan, under such a pressure that the boiling shall take place at a temperature not exceeding 150° F. (65·5° C.) It is lastly run into moulds and allowed to cool, when it becomes perfectly solid, and is then ready for the market. On the Continent this variety of sugar is also made from potatoes, the process of conversion being essentially the same.

The second mode of manufacture is to treat 100 parts of grain, or other substance containing starch, with 40 parts of water and 3 parts of sulphuric acid. This mixture is allowed to stand for twenty-four hours in the cold, and then heated for about three minutes in the converter under a pressure of 65 lb. of steam. The removal of the sulphuric acid, the filtering and concentration, are then effected as before described; but by this process the conversion of the starch into glucose is not so complete as by that first given.

Maltose is the name given to an article prepared from starch by the action of malt-extract. It differs in character from dextro-glucose in its optical properties, having a dextro-rotatory power of 150°, that of dextro-glucose being 56°. It also differs in its cupric-oxide reducing power, which is about 62 per cent. of that of glucose. Dubrunfaut, the French chemist, was the first to regard this as a distinct sugar, but we are indebted largely to the labours of O'Sullivan for what we know in regard to it. Previous to the researches of these chemists it was considered to be a mixture of dextro-glucose and dextrin, and some of its chemical reactions favoured this view of its constitution.

Lævulose or Lævoglucose is a colourless uncrystallizable syrup as sweet as that of cane-sugar, and contained in equal proportions with dextro-glucose in invert-sugar. Its distinctive property is the power it possesses of turning the plane of polarisation to the left, and hence its name. This rotatory power, unlike that of dextro-glucose, varies very considerably with the temperature, being 106° at 57° F. (14° C.) and only 53° at 196° F. (91° C.) It is formed along with dextro-glucose by warming a solution of cane-sugar with dilute acid; by leaving it exposed to the air for a time; or by exposing it to the action of yeast. It exists with dextro-glucose in honey and many fruits, and is also found in treacle and molasses, which, as already stated, are mixtures of invert and cane sugars.

Mannitose is the sugar derived from manna, from several sea-weeds, and from mushrooms. To prepare it, a substance called Mannite is dissolved out of manna by boiling alcohol. This is oxidised, in presence of platinum black, into mannitic acid and mannitose, and the mannitic acid is removed by means of lime, with which it forms a compound insoluble in alcohol.

Besides all these there is another sugar derived from milk, named "Lactose," which, however, will be more fully referred to under the head of Milk.

History.—The word *sugar* is probably derived from the Sanscrit word *sarkara*, which in the Persian became *shukkur*. It is supposed by Humboldt to have been known to the Chinese in very early times, and it is not improbable that it was in use by the ancient Jews, and that the Hebrew word occurring frequently in the Old Testament, sometimes rendered "calamus," and sometimes "sweet cane," has reference to it. Sugar is mentioned occasionally by the early historians as "honey from reeds," "saccharon," etc., and many medicinal properties are by them ascribed to it. Its introduction into Europe for dietetic purposes was in some measure due to the Crusaders. There can be little doubt but that the sugar-cane is natural to many islands in the West Indies, as well as to parts of South America, but it appears to be equally clear that the natives were not acquainted with the means of extracting sugar from it until after the Spanish and Portuguese colonisation. So quickly, however, did the manufacture spread that, according to the testimony of Oviedo, no fewer than thirty sugar-mills were established in Hispaniola (St. Domingo), in 1535; and in the island of St. Thomas (Portuguese), in 1620, there were seventy works, each employing not fewer than 200 slaves, and forty ships being loaded yearly with the produce. We find that, in 1726, the French produced 33,000 hogsheads of sugar in St. Domingo, in 1742, 70,666 hogsheads; and in Martinique, Guadaloupe, and the lesser isles, 51,875 hogsheads. The whole produce of the British West India Islands imported into Great Britain was 60,950 hogsheads, but at this time the East Indies had been opened out by the enterprise of the East India Companies, and we were obtaining a large supply from that quarter of the globe.

The art of refining sugar began to be practised in England in 1544.

The consumption of sugar in Great Britain in 1700 was estimated, according to McCulloch, at 10,000 tons; in 1790 it had reached 81,000 tons; in 1808, although the duty had been gradually raised from 3s. 5d. per cwt. (the rate during Queen Anne's reign) to 27s., the consumption had increased to 142,000 tons. It fell during the Napoleon wars to 100,000 tons, and from that time there has been a steady increase, the amount in 1867 being 546,000 tons. The imports in 1877, including beet-sugar, were 1,003,161 tons; and in 1880, 1,001,285 tons.

The discovery that beetroot contained sugar identical with that obtained from the sugar-cane, was first made known by Margraf, in 1747. The culture of the beet for the purpose of sugar manufacture did not, however, make much progress during that century, and it was mainly in consequence of the policy of Napoleon that it could be said to have come into competition with the product of the sugar-cane. From the peace of 1815 to the year 1860 it passed through many vicissitudes; and, even after the latter date, it was confidently affirmed by refiners in this country that beet-sugar could not be deprived of the peculiar flavour attaching to beetroot itself, and on that account would never be a dangerous rival in this country to the product of the sugar-cane. Difficulties which were then considered insuperable have, however, since been overcome, and the purest loaf-sugar is now found in our markets prepared from beet. In 1858 Johnston calculated the total beetroot sugar manufactured in the world at 159,821 tons. In 1869 it was estimated at 689,500 tons for Europe alone; in 1877-78, at 1,420,800 tons; in 1878-79, at 1,574,100 tons; in 1879-80—there being a failure in the crops—1,275,000 tons.

The production of glucose from starch was first accomplished by M. Kirchoff, of St. Petersburg, in 1702. He recommended boiling the starch with 1 per cent. of sulphuric acid for thirty-six hours. The manufacture had been carried on for some years on the Continent before its introduction into this country, and a liquid form of it is imported and used to some extent by confectioners. The solid form is, however, chiefly used by brewers, and comes into direct competition with the glucose made in this country by one of the methods we have described.

CHEMICAL COMPOSITION.

Pure cane-sugar has been analysed by many chemists. One of these results is as follows:

Carbon	42·15 per cent.
Hydrogen	6·47 ,,
Oxygen	51·38 ,,

This analysis agrees very closely with the recognised formula for sugar, $C_{12}H_{22}O_{11}$. The specific gravity of the crystals of pure cane-sugar, according to Joule and Playfair, is 1·593 at 39° F. (4° C.) It melts at about 320° F. (160° C.) without losing weight to a clear pale yellow liquid, which is a mixture of dextro-glucose and levolusan.

$$C_{12}H_{22}O_{11} = \underset{\text{Dextro-glucose.}}{C_6H_{12}O_6} + \underset{\text{Levolusan.}}{C_6H_{10}O_5}.$$

Heated to 410° F. (210° C.) sugar is converted into a dark brown substance called Caramel, which is extensively used for colouring liquids.

Cane-sugar, unlike glucose, is decomposed by strong sulphuric acid, with copious formation of carbonaceous matter. It is not turned brown when treated with alkalis, and is insoluble in cold absolute alcohol. It dissolves in one-third of its weight of water at mean temperature, and in all proportions in boiling water. From a solution containing 5 parts of sugar to 1 part of water, three-fifths of the sugar crystallizes on cooling in four or six sided rhomboidal prisms.

ESTIMATION OF SUGAR.

Cane-sugar does not precipitate the suboxide of copper from alkaline solutions of cupric tartrate, but it is very readily converted by boiling with dilute acid into invert-sugar, which does possess that property. Advantage is taken of this in what is generally called "Fehling's test." A solution is made by dissolving 86 grams of tartaric acid in crystals with 104 grams of caustic soda. To this is added 29 grams of sulphate of copper dissolved in water. The bulk is then made up by additional water to 1 litre. This is Fehling's solution, and in its application for the estimation of sugar it may be used either "volumetrically" or "gravimetrically:" in either case it is necessary in the first place to have a standard. In the volumetric process, which is the easier, ·625 gram of pure cane-sugar is for this purpose boiled for ten minutes with about 4 ounces of water acidulated with 5 drops of concentrated sulphuric acid. The solution is then cooled, neutralised with solution of caustic soda, and made up to a bulk of 250 cubic centimetres. Twenty-five cubic centimetres of the copper solution are then heated in a white glass flask to the boiling point, and the sugar solution is run into it from a burette, care being taken not to add more than will reduce the whole of the copper. It will generally be found that 40 cubic centimetres of the sugar solution, which correspond to ·1 gram cane-sugar, or ·105 gram glucose, will be required to reduce the copper or decolourise 25 cubic centimetres of copper solution. If more or less than

40 cubic centimetres are required, a corresponding difference will have to be made in the quantities of cane-sugar and glucose represented respectively. This result is applied in the examination of saccharine substances or solutions in the following way : If a known weight—say 8 grams—of a liquid which contains glucose and cane-sugar be taken and made up to 250 cubic centimetres, and if it be found that 45 cubic centimetres of this diluted solution are required to reduce the copper in 25 cubic centimetres of Fehling's solution, the percentage of glucose is thus found :

$$\frac{250 \times 100 \times \cdot 1}{45 \times 8} = 6\cdot 94 \text{ per cent., the cane-sugar equivalent, or } 7\cdot 30 \text{ per cent. glucose.}$$

It is then necessary to make a second experiment to find the total amount of sugar present. A less weight than before—say 4 grams—is taken and boiled for four minutes with about 4 ounces of water and 5 cubic centimetres of normal sulphuric acid to invert the cane-sugar. It is then neutralised with soda and made up as before to 250 cubic centimetres at 60° F. (15·5° C.), and if it be then found that 50 cubic centimetres of this solution are necessary to reduce the copper in 25 cubic centimetres of Fehling's solution, the total sugar in the liquid, calculated as cane-sugar, is as follows :

$$\frac{250 \times 100 \times \cdot 1}{50 \times 4} = 12\cdot 5 \text{ ; and } 12\cdot 5 - 6\cdot 94 = 5\cdot 56, \text{ the percentage of cane-sugar present.}$$

In the gravimetric method, the standard is the quantity of cuprous oxide precipitated by a given quantity of sugar-solution. The cuprous oxide being either ignited with a little nitric acid, and weighed as cupric oxide, or, as recommended by Pavey, the suboxide of copper is dissolved, and the copper precipitated from it by electrolysis and weighed. Fehling's test, although fairly accurate, where the percentage exceeds 0·5 per cent., is not well adapted for cases in which it falls below that quantity. Knapp's method, based upon the decomposition of an alkaline solution of cyanide of mercury, has been suggested where the quantity of sugar is very small. The standard liquid is prepared by dissolving 10 grams pure dry mercuric cyanide in water, adding 100 cubic centimetres of sodium hydrate solution—specific gravity 1·145, and diluting to 1,000 cubic centimetres. Ten cubic centimetres of this solution are equal to 25 milligrams of glucose. To apply the test, 10 cubic centimetres of the solution, diluted with from 20 to 30 cubic centimetres of water, are heated to the boiling point. The sugar solution is run in from a burette until the whole of the mercury is precipitated. When the precipitate has settled, a drop of the supernatant liquid, which has a more or less yellow tint, is transferred by means of a capillary tube to a thin pure white Swedish filter-paper. This paper is held, first over a bottle containing strong hydrochloric acid, and then over a saturated sulphuretted hydrogen solution. The slightest trace of mercury is shown by the production of a light brown or yellow stain. It is well to place a drop of the original liquid beside that which has been subjected to the action of hydrochloric acid and sulphuretted hydrogen for comparison.

Cane-sugar, when exposed to the action of yeast, is rapidly changed, first into invert-sugar, and then into alcohol and carbonic dioxide, the following being the reactions:

1st. $C_{12}H_{22}O_{11} + H_2O = 2C_6H_{12}O_6$.
2nd. $C_6H_{12}O_6 = 2CO_2 + 2C_2H_6O$.

A process based upon the quantity of alcohol produced by fermentation from a given quantity of sugar has been long in use for estimating the percentage of sugar in substances to which, owing either to their colour or to the fact that they contain matter other than sugar capable of reducing salts of copper, Fehling's method cannot be applied. In determining the amount of sugar by the fermentation process, the quantity taken, in order to insure complete fermentation, should not exceed 100 grains. Assuming that 100 grains of the sample to be analysed, when dissolved in about a quart of water and fermented with 200 grains of pressed yeast, yield a distillate of 1,000 fluid grains of a density of 990·3 at 60° F. (15·5° C.), and that 200 grains of yeast similarly fermented yield a distillate of the same volume, having a density of 998·3, the following calculations will give the percentage of cane-sugar or glucose present. By Gilpin's tables it will be found that mixtures of alcohol and water of a density of 990·3 and 998·3 contain respectively 5·52 grains and ·88 grain by weight of absolute alcohol in each 100 fluid grains, and therefore 55·2 and 8·8 grains respectively will be contained in each distillate. Deducting the latter from the former, there remain 46·4 grains of absolute alcohol as having been produced from the sugar. By the equations given above 342 parts by weight of cane-sugar, or 360 parts of glucose, are seen to be necessary to produce theoretically 184 parts of absolute alcohol; hence

$$\frac{46\cdot4 \times 342}{184} = 86\cdot2$$

is the percentage of cane-sugar; or

$$\frac{46\cdot4 \times 360}{184} = 90\cdot7$$

the percentage of glucose in the substance analysed.

Should the rectifying power of the distilling apparatus used not be sufficient to insure the collection of the whole of the alcohol in the first distillate of 1,000 fluid grains, a second similar bulk must be distilled over, and the amount of alcohol found added to that obtained in the first distillate.

As small quantities of glycerin and succinic acid are formed during fermentation, the amount of sugar, calculated from the alcohol produced, is invariably less than the true quantity, even under the most favourable conditions. It is sometimes desirable, therefore, in practice, to make experiments with pure cane-sugar, and to use the highest alcoholic result obtained as a factor for calculating the amount of sugar contained in saccharine substances submitted to the fermentation test.

Another process is to estimate the sugar from the loss of carbonic dioxide in the course of fermentation. Its application requires very great care in manipulation, and it is not likely to be resorted to, except in cases in which the other processes are inapplicable.

The most ready method of estimating the percentage of pure cane-sugar in raw sugars, whether derived from beet or cane, is by the polariscope. The principle of this instrument is that a plane polarised ray of light may always be considered as made up of two circularly polarised rays, and if these pass through a medium, such as sugar, tartaric acid, etc., which retards the one more than the other, the plane of polarisation of their resultant when they leave the medium will in general not be the same as that of the incident ray—or, in other words, it will have been caused to rotate through a certain angle, sometimes to the right and sometimes to the left. This rotation varies in the different descriptions of sugar, both in regard to the angle and the direction. If a tube 1 decimetre long be filled with a solution of pure cane-sugar, containing 1 gram in every cubic centimetre of fluid, it will rotate the plane of polarisation 73·8 degrees to the right, and this is called the specific rotatory power of pure cane-sugar. Rotation is in proportion to the length of the tube, and the mass of substance possessing the rotatory power, water being quite neutral. It follows, therefore, that if we take a solution, containing a decigram of pure cane-sugar in every cubic centimetre of fluid, the tube being the same length as before, we obtain a rotation of 7·38°. If we then take an impure cane-sugar, and make a solution such that it shall contain 1 decigram in every cubic centimetre of liquid, fill a tube, 1 decimetre in length, with such solution, and find the rotation to be 6·3°, we should, supposing no invert-sugar to be present, find the percentage of sugar by the following proportion: as 7·38 : 6·3 :: 100 : x. The rule for finding the specific rotation from the observed rotation is: Divide the observed rotation by the length of the tube, multiplied by the weight of sugar in each cubic centimetre of liquid, 1 gram being the unit of weight, and 1 decimetre the unit of length. Thus if a solution, containing 0·150 gram of sugar in every cubic centimetre of fluid, has an observed rotatory power of 16° in a tube 2 decimetres long, the specific rotatory power would be

$$(1) \quad \frac{16}{2 \times 0\cdot 150} = 53\cdot 33°$$

and if this were a cane-syrup, the percentage of sugar would be as $73\cdot 8 : 53\cdot 33 :: 100 : x$. But raw sugars generally contain more or less invert-sugar; and as glucose has a specific rotatory power of 56° to the right, while lævulose, at a temperature of 57·2° F. (14° C.), rotates 106° to the left, the specific rotation of invert-sugar at 57·2° F. must consequently be

$$(2) \quad \frac{106 - 56}{2} = 25°$$

to the left. If, therefore, at the temperature of 57·2° F. (14° C.), we obtain a solution of sugar which produces a specific rotatory power of 67°, and we find by Fehling's test that it contains 4 per cent. of invert-sugar, we have the data necessary for estimating the cane-sugar. Let a = the percentage of invert-sugar by Fehling's test, b the specific rotatory power of the sugar examined, and x the percentage of crystallized cane-sugar,

$$(3) \quad \text{Then } \frac{100b + 25a}{73\cdot 8} = x.$$

The polariscopes now in most general use are those in which the scale, instead of being marked with the angle, has upon it the percentage of crystallized sugar corresponding to that angle, the quantity of sugar used, and the volume to which it is made up, the length of tube being always the same.

In the trade the percentage of crystallized sugar is not regarded as the sole criterion of value. The percentage corresponding to the angle, given by the mixed sugars, is what is called by sugar merchants the percentage of crystallized sugar; and the percentage of ash, as well as the appearance of the sugar, is taken into account along with this indication in fixing the price.

It should also be remarked that beet-sugar, as will be seen by the analyses of various samples, contains very little invert-sugar, so little indeed that it is disregarded on the Continent.

The following table exhibits the specific rotatory power of commercial sugars, with the amount of cane-sugar calculated directly therefrom, and also the quantities of invert-sugar, ash, and moisture:

Description of Sugar.	Specific Rotatory Power.	Cane-sugar corresponding to Rotation.	Invert-Sugar.	Ash.	Moisture.
Beet	67·6°	91·6	Trace	2·75	2·73
,,	68·0°	92·1	,,	1·93	2·82
,,	70·8°	95·9	,,	1·35	2·29
,,	71·0°	96·2	,,	1·22	2·06
,,	69·9°	94·8	,,	0·75	1·68
,,	69·7°	94·5	,,	1·42	2·74
,,	66·7°	90·4	,,	2·30	4·04
China	59·6°	80·8	6·94	2·93	5·58
Natal	61·8°	83·7	3·38	2·89	6·86
Jaggery	60·1°	81·4	3·06	6·03	5·22
Trinidad	65·2°	88·3	5·00	1·19	4·64
Havannah	69·6°	94·3	1·18	0·36	2·28
Mauritius	66·5°	90·1	3·25	1·00	2·78
Manilla	61·1°	82·8	7·29	3·28	4·76
Egyptian	60·6°	82·1	4·81	3·70	3·00
,,	67·0°	90·8	3·47	—	—

If the formula (3) given above be applied to these results, the proportion of cane-sugar indicated will be from 1 to 2 per cent. more than the amount calculated directly from the angle of rotation.

It is sometimes useful to determine the approximate value of a sugar from the specific gravity of its aqueous solution; and for this purpose the following table, which shows the specific gravities

SUGAR.

of solutions containing from 1 to 66 per cent. by weight of pure cane-sugar, has been prepared:

Parts by Weight.		Specific Gravity of the resulting Solution at 60° F. (15·5 C.)	Parts by Weight.		Specific Gravity of the resulting Solution at 60° F. (15·5 C.)
Sugar.	Water.		Sugar.	Water.	
1	99	1003·89	34	66	1149·31
2	98	1007·82	35	65	1154·28
3	97	1011·77	36	64	1159·28
4	96	1015·75	37	63	1164·32
5	95	1019·76	38	62	1169·40
6	94	1023·79	39	61	1174·51
7	93	1027·84	40	60	1179·66
8	92	1031·93	41	59	1184·83
9	91	1036·06	42	58	1190·05
10	90	1040·21	43	57	1195·31
11	89	1044·39	44	56	1200·60
12	88	1048·61	45	55	1205·93
13	87	1052·86	46	54	1211·29
14	86	1057·13	47	53	1216·69
15	85	1061·44	48	52	1222·14
16	84	1065·78	49	51	1227·61
17	83	1070·14	50	50	1233·13
18	82	1074·54	51	49	1238·69
19	81	1078·97	52	48	1244·28
20	80	1083·43	53	47	1249·92
21	79	1087·93	54	46	1255·59
22	78	1092·45	55	45	1261·30
23	77	1097·01	56	44	1267·04
24	76	1101·60	57	43	1272·83
25	75	1106·22	58	42	1278·65
26	74	1110·87	59	41	1284·51
27	73	1115·56	60	40	1290·42
28	72	1120·28	61	39	1296·35
29	71	1125·04	62	38	1302·33
30	70	1129·83	63	37	1308·35
31	69	1134·64	64	36	1314·40
32	68	1139·50	65	35	1320·50
33	67	1144·39	66	34	1326·64

ADULTERATION.

Owing probably to the low price of cane-sugar, and to the difficulty of finding a suitable cheap adulterant, it is remarkably free from sophistication. If glucose or starch sugar were suspected to have been added, the quantity present might be estimated by Fehling's test, or by the polariscope, according to the methods previously described.

Perhaps the most serious deceit now practised upon the consumer of sugar is the sale of the lower products of the refiner as raw sugar. These products, technically known as "pieces," are caused to crystallize in very small crystals, and thus to hold a comparatively large percentage of water as well as of invert-sugar. They possess much less sweetening power than raw sugar, but having generally less colour are erroneously supposed by the public to combine cheapness with superiority of quality.

HONEY.

Honey, as is well known, is the saccharine substance collected by bees from the nectaries of flowers, and stored by them in combs for winter use. It consists, as might be expected from its origin, of a mixture of various bodies, the principal of which are dextroglucose, lævoglucose, and a third body, which is probably one of the less known sugars. Besides these there are small proportions of wax, gum, pollen, and other vegetable and some mineral matters.

The odour and flavour of honey vary according to the nature of the plants from which it has been collected. When new, it flows freely from the comb, and crystallizes after a time into a semi-solid mass. This change takes place to some extent in the comb if left for several months, and then heat and pressure are required for its removal. It is probable that the saccharine substances extracted from the flowers undergo modification in the honey-bag of the bees. In connection with this it has been observed that bees fed upon a solution of pure cane-sugar readily produce wax therefrom for the formation of the comb.

The third principal constituent referred to as probably one of the less known sugars is only partially fermentable, and has no direct action upon cupric tartrate, but is gradually converted into glucose when boiled for several hours with a few drops of sulphuric acid. Some chemists have represented this body as cane-sugar, but when the samples of honey mentioned in the table below were boiled with sulphuric acid for a sufficient time to invert cane-sugar, the additional reduction of cupric tartrate corresponded to only 2·10 per cent. of cane-sugar in the honey from comb, and to less than 1 per cent. in the other four samples. It is therefore evident that but a small part of it, if any, is cane-sugar, and we have therefore placed it in the table as "sugar not identified." As honey is acid, and undergoes a slight fermentation, cane-sugar, even if originally present, would gradually be transformed into invert-sugar, and thus escape detection.

The following results have been obtained from an examination of five samples of commercial honey:

Percentage of	Taken from Comb.	Californian.	Narbonne.	West Indian.	Transylvanian.
Moisture	17·42	23·32	17·10	19·65	22·75
Glucoses	71·66	68·52	74·04	69·34	66·57
Sugar not identified	10·12	4·48	7·10	7·55	7·97
Gum	·23	·17	·13	·36	·22
Inorganic Matter	·13	·49	·28	·27	·32
Wax, Pollen, etc., and Loss	·44	3·02	1·35	2·83	2·17
Total	100·	100·	100·	100·	100·

The adulterants said to have been found in honey are gypsum, chalk, pipeclay, starch, glucose, and cane-sugar, but at the present day the three former are not likely to be used. Starch may be readily found by the microscope and solutions of iodine. Glucose cannot be detected by chemical means, and only by the polariscope when in sufficient quantity to change the angle of rotation beyond the limits found in genuine honey. Cane-sugar may be found by the copper test, and also by the polariscope; and in this case the readings are taken both before and after inversion, the difference in the readings being proportionate to the amount of cane-sugar present.

APPENDIX.

TABLE I.

For Converting Degrees of the Centigrade Thermometer into Degrees of Fahrenheit's Scale.

Centigrade.	Fahrenheit.	Centigrade.	Fahrenheit.
0°	32°	105°	221°
5	41	110	230
10	50	115	239
15	59	120	248
20	68	125	257
25	77	130	266
30	86	135	275
35	95	140	284
40	104	145	293
45	113	150	302
50	122	155	311
55	131	160	320
60	140	165	329
65	149	170	338
70	158	175	347
75	167	180	356
80	176	185	365
85	185	190	374
90	194	195	383
95	203	200	392
100	212	205	401

TABLE II.

COMPARISON OF FRENCH AND ENGLISH MEASURES OF WEIGHT.

		Grains.
Milligram	=	0·015432
Centigram	=	0·154323
Decigram	=	1·543235
Gram	=	15·432349
Decagram	=	154·323488
Hectogram	=	1543·234880
Kilogram	=	15432·348800
Myriagram	=	154323·488000

INDEX.

	PAGE
Acorn	67
Adulteration of cocoa	88
,, coffee	50
,, honey	116
,, sugar	114
,, tea	22
Analysis of cocoa	82
,, coffee	48
,, honey	116
,, tea	16
Arrowroot	92
Bean	64
Botanical origin of cocoa	72
,, ,, coffee	40
,, ,, tea	1
Carrot	63
Chemical composition of cocoa	76
,, ,, coffee	42
,, ,, honey	116
,, ,, sugar	105
,, ,, tea	5
Chicory	60
Cocoa	72
,, adulteration	88
,, analysis	82
,, botanical origin	72
,, chemical composition	76

	PAGE
Cocoa description	73
,, history	75
,, microscopic structure	81
Coffee	40
,, adulteration	50
,, analysis	48
,, botanical origin	40
,, chemical composition	42
,, description	40
,, history	41
,, microscopic examination	59
,, miscroscopic structure	46
Date coffee	71
,, stone	70
Description of cocoa	73
,, coffee	40
,, honey	115
,, sugar	98
,, tea	1
Elder leaf	34
Estimation of cane-sugar	106
,, glucose	106
Fig	69
Glucose, manufacture of	101
,, estimation of	106

INDEX.

	PAGE
History of cocoa	75
,, coffee	41
,, sugar	103
,, tea	3
Honey	115
Lævulose	102
Locust bean	65
Leaf of elder	34
,, sloe	38
,, tea	12
,, willow	36
Maltose	102
Mannitose	103
Mangold wurzel	61
Microscopic examination of cocoa	81
Microscopic examination of coffee	59
Microscopic examination of tea	32
Microscopic structure of acorn	68
,, arrowroot	92
,, bean	64
,, carrot	63
,, chicory	60
,, cocoa	81
,, coffee	46
,, date stone	70
,, elder leaf	35
,, fig	69
,, locust bean	65
,, mangold wurzel	61
,, pea	65
,, potato	95
,, sago	93
,, sloe leaf	39
,, tea leaf	14

	PAGE
Micro. structure of tous les mois	95
,, turnip	62
,, willow leaf	37
Nitrogen in albumin	7
,, tea	7
Parsnip	63
Pea	65
Polariscopic examination of sugar	110
Potato	95
Sago	93
Sloe	38
Starch of arrowroot	92
,, potato	95
,, sago	93
,, tous les mois	95
Sugar	97
,, adulteration	114
,, chemical composition	105
,, description	98
,, estimation	106
,, history	103
,, origin	97
,, polariscopic examination	110
Tea	1
,, adulteration	22
,, analysis	16
,, botanical origin	1
,, chemical composition	5
,, description	1
,, history	3
,, microscopic examination	32
,, ,, structure	12
Tous les mois	95
Turnip	62

THE END.